THE
AQUATIC
APE

Also by Elaine Morgan:

THE DESCENT OF WOMAN
FALLING APART

THE AQUATIC APE

Elaine Morgan

𝔰𝔇

STEIN AND DAY/Publishers/New York

To Morien

First published in the United States of America in 1982
Copyright © 1982 by Elaine Morgan
All rights reserved
Printed in the United States of America

STEIN AND DAY/*Publishers*
Scarborough House
Briarcliff Manor, N.Y. 10510

Library of Congress Cataloging in Publication Data

Morgan, Elaine.
  The aquatic ape.

  Bibliography: p.
  Includes index.
  1. Human evolution.   I. Title.
GN281.4.M67   1982b   573.2   82-40012
ISBN 0-8128-2873-9

Acknowledgments

I am indebted to the many people who helped to make this book possible—primarily to Sir Alister Hardy for his active support and encouragement and to Leon P. La Lumiere, Jr. for permission to reprint extracts from his paper *The Evolution of Human Bipedalism.*

I would like to thank R. D. Martin for his advice on various aspects of primate biology. As a former student of Alister Hardy and a great admirer of his pioneering work in marine biology, he was glad to provide comments.

Thanks are also due to the following:

Morien Morgan for his support and collaboration throughout; Chuck Milliken for collecting material and professional opinions and for drawing the charts; Michael Welpley for the drawings; Dylan Morgan for comments and ideas; Illustration Research Service for supplying photographs (Plates 5, 6, 7 and 8); Angus & Robertson (UK) Ltd for permission to incorporate in the graphs (pages 70-72) data from *The Ocean World of Jacques Cousteau—Mammals in the Sea.*

For the photographs: Chris Gregory (Plate 2); Bruce Colman Ltd (Plate 5); Zoological Society of London (Plate 3); Queen Mary's Hospital Roehampton (Plate 4); Syndication International Ltd (Plates 7 and 8); Anita Corbin and the *Sunday Times* (Plate 6); and Camera Press (Plate 1).

And to Christopher C. Blake, Sheldon Cholst, S. C. Cunnane, Erika von Herbst, and hundreds of other correspondents from all over the world who have submitted observations, queries, references, press cuttings, arguments, suggestions, and moral support since the publication of *The Descent of Woman* in 1972.

Contents

I
The Emergence
of Man

1 Something Happened

In 1871 Charles Darwin published *The Descent of Man,* proposing that man and apes are descended from a common ancestor.

No anthropologist today questions his basic premise. There is total agreement about how to explain the similarities between men and apes.

The impression is sometimes given that there is an equal consensus on how to explain the differences between them. This impression is misleading.

Considering the very close genetic relationship that has been established by comparison of biochemical properties of blood proteins, protein structure and DNA, and immunological responses, the differences between a man and a chimpanzee are more astonishing than the resemblances. They include structural differences in the skeleton, the muscles, the skin, and the brain; differences in posture associated with a unique method of locomotion; differences in social organization; and finally the acquisition of speech and tool using, together with the dramatic increase in intellectual ability that has led scientists to name their own species *Homo sapiens sapiens*—wise wise man.

During the period when these remarkable evolutionary changes were taking place, other closely related apelike species changed only very slowly and with far less remarkable results. It is hard to resist the conclusion that something must have happened to the ancestors of *Homo sapiens* which did not happen to the ancestors of gorillas and chimpanzees.

Finding an answer to the question "What happened?" is made more difficult by the fact that no fossil relics have yet been discovered from the period when these changes were taking place. Fossils exist of an apelike creature, a possible remote human ancestor known as *Ramapithecus*, dating from around 9 million years ago. From about 3½ million years ago we have reliable fossils—and even fossil footprints—of a creature who walked upright on two legs. Between these two dates comes the gap in the fossil record which Richard Leakey aptly described as "a yawning void." And it was during this blank period that man's ancestors apparently embarked on the divergent evolutionary path leading to their separation. Whatever happened, happened then.

In the absence of direct evidence, the only way of deducing what did happen is by inference from: (a) what we know of apes; (b) what we know of man; (c) what we know of the fossils; (d) what we know of conditions in Africa at the relevant period; and (e) what we know of evolutionary processes in general.

Using these methods and arguing from agreed data, different people have nevertheless arrived at different conclusions as to what was the major factor causing the forerunners of man to diverge sharply from their anthropoid relatives.

There are three main schools of thought. For the purposes of this book they will be referred to as (1) the savannah theory; (2) the neoteny theory; (3) the aquatic theory.

The theories are in no way mutually exclusive. Yet they diverge on some of the key questions concerning unique features of human physiology—such as "Why are men's bodies less hairy than the bodies of apes?" or "Why are their skulls different?"

2 The Savannah Theory

The savannah theory postulates that the evolution from ape to hominid proceeded in a smooth regular line throughout the "fossil gap." Despite the fact that we have yet to find remains from these intermediate stages of development, the adherents of this theory consider that it is only a matter of time before they come to light.

This theory argues that two major factors contributed to the accelerated rate of evolutionary change—one climatic, one behavioral. The climatic change resulted in a dwindling of the forested areas of the African continent so that large areas became covered with grass and scrub. The hominids, it is claimed, are descended from those apes who left the trees and moved out onto the grassy plains or savannahs. Gorillas and chimpanzees are descended from the ones who remained in the trees.

The behavioral change was one of diet. Forest-dwelling apes are not normally troubled by food shortages—they are vegetarians surrounded by plentiful year-round supplies of fruit and lush vegetation. These would have been scarcer on the savannah, so the apes began to vary their diet. Initially they did this by catching small game or, possibly, by scavenging the remains of kills made by the larger carnivores. Thus by degrees they turned themselves into meat-eaters, and finally hunters. It is known that male chimpanzees, though largely vegetarian, occasionally hunt for meat. (The females seem to obtain their required animal protein through eating small arthropods such as termites.)

According to this line of reasoning, each of the major evolutionary modifications leading from ape to man is a direct or indirect result of becoming a plains dweller and a hunter. Thus, the ape learned to stand up straight in order to see further over the plains searching for prey; and he learned to run fast on two legs in order to pursue game while leaving his hands free to carry weapons.

As a forest dweller he had been accustomed to a leisurely and well shaded life; so when he ran after his prey in the sunshine, he was liable to get overheated. Therefore he gradually discarded most of his body hair in order to keep cool.

The necessity of fashioning weapons to kill other animals, and tools to skin and butcher them, sharpened his intelligence; he developed a larger cranium because he needed a bigger brain.

The hunters' need to cooperate in the chase led to the evolution of speech; their need to return to a communal base with their kill led to a more organized social life, pair bonding, divisions of labor, and other recognizably human activities.

3 The Neoteny Theory

Neoteny is a phenomenon that occurs repeatedly in the evolutionary history of various species. A classic example of neoteny is that of a Mexican salamander, a newtlike amphibian called *Ambystoma*, much valued in Mexico City as a table delicacy.

An adult salamander, after a larval stage corresponding to the tadpole stage of a frog, loses its gills and emerges from the water as an air-breathing, land-dwelling, four legged animal.

Sometimes, however, the metamorphosis from tadpole to salamander fails to take place. The immature salamander remains in the water, retains all its larval characteristics (the external gills, the lidless eyes, the teeth in both jaws) and in that condition mates and reproduces its kind without ever attaining the normal salamander condition of adulthood.

This is assumed to happen because under certain conditions the larval form is better fitted to survive than the adult one would be. For example, an adult salamander can survive only if its skin remains moist, and a prolonged heat wave might dry out the damp woodlands that are its normal habitat.

Neoteny, then, is an evolutionary trick by which an animal retains throughout its life features that in its ancestors were typical of an immature stage of existence—sometimes a very early or even fetal stage.

Figure 1 vividly illustrates one reason why the concept of neoteny has been applied to human evolution. The face of the adult chimpanzee does not bear a very strong likeness to a human face, but in the case of the infant chimpanzee the resemblance is striking. It is possible, then, to regard man not as the hunting ape but as the neotenic ape—an ape which has become paedomorphic (child-shaped).

As already described, some animals become paedomorphic by omit-

Fig. 1.   Baby and adult chimpanzee (after Naef, 1926 b). In the case of the
infant chimpanzee, the resemblance to man is striking.

ting the last step in their metamorphosis. Man is believed to have
become paedomorphic by a different method—by retardation, or the
slowing down of all the stages in his development. This could possibly
explain why man takes such a long time to grow up—much longer than
other apes—and also why he lives much longer than they do.

Other unique human features are held to be the consequence of the
same process. For example, it is argued that human beings are comparatively hairless because the fetus of every ape is at one stage hairless and in
Homo sapiens this feature has been retained into adulthood. Similarly,
all ape fetuses have larger skulls than adult apes relative to their body
size—and the larger skull is one of the features distinguishing men from
apes.

Strictly speaking, neoteny is not an explanation of evolutionary
change—it is only a mechanism by which such changes may be brought
about. In practice, however, larger claims are made for it. Alternative
theories are dismissed on the grounds that they are unnecessary, since
neoteny supplies so many of the answers. It will be more convenient,
therefore, to treat it as a theory in its own right.

4 The Aquatic Theory

The aquatic theory starts out with the observation that among those
morphological and physiological features commonly regarded as being
unique to man, a surprising number are not really unique at all. They
may be unique among land mammals, but they are quite common—in
some cases practically universal—among those species of mammals
which have left the land and returned to an aquatic existence.

Time after time throughout evolutionary history this process has
taken place. Reptiles, birds and mammals that had become fully adapted
to life on land and begun breathing with lungs have abandoned their
terrestrial existence, gone into the water, and become modified in
various ways.

This process is very ancient. Even before the first mammals came into
existence, a four-legged, air-breathing, land-dwelling dinosaur went into
the sea and remained long enough to turn its legs into flippers and itself
into an ichthyosaur ("fish lizard") before becoming extinct. The ichthyosaur is only one example of many early reptiles that became aquatic.
The pleiosaur, the mesosaur, the aigialosaur, the dolichosaur, and others
followed the same path.

Among mammals, the first to return to the water, some 70 million
years ago, were the cetaceans (the whales, dolphins, and porpoises). Like
all mammals they are warm-blooded, breathe air, give birth to live
offspring, and suckle their young. Their skeletons still exhibit in modified form the standard mammal structure—with spinal column, fore-
limbs modified into flippers, and the vestiges of the pelvic girdle to
which hind legs were formerly attached. But they have lost all their hair
and, in their general shape and mode of life, have grown to resemble fish
so closely that Catholics used to be allowed to eat them on Fridays.
Although cetaceans are classed as a single order, it is quite possible that
they are descended from two or three different mammalian species. It
has been suggested that each of these species returned separately to the
sea, and that they only grew to resemble one another over a long period
of time because of the pressures of their new environment.

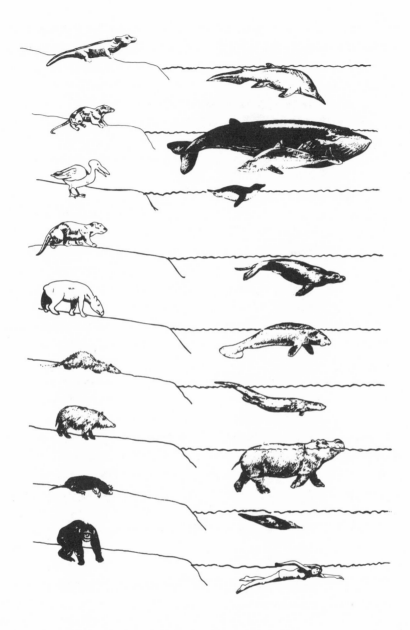

Fig. 2. Time after time throughout evolutionary history, this process has taken place.

The next to go into the sea, more than 50 million years ago, were vegetarian hoofed mammals related to the elephant. Their descendants are the sirenians, or sea cows. The dugong and the manatee are still extant though their numbers are greatly diminished. The largest of the sirenians, the North Pacific *Hydrodamalis* (Steller's sea cow, or rhytina) became extinct in the eighthteenth century.

Between 25 and 30 million years ago some bearlike carnivorous mammals took to the water. These were the ancestors of the present-day fur seals, sea lions and walruses. Around the same time, the ancestor of the true seals left the land. This was another carnivore, possibly resembling a dog.

In the view of J. Z. Young: "These returned aquatics are especially interesting because each type when it first re-enters the water seems to be not very well suited to that medium." Similarly, the shrewlike creature that began to turn into a bat cannot at first have been very well suited to life in the air. Despite the improbability of such dramatic changes in lifestyle, the fact remains that the adoption of aquatic habits happened over and over again.

There are aquatic birds (penguin), aquatic carnivores (sea lion, seal, and otter), aquatic insectivores (water shrew, desman), aquatic hoofed mammals (hippopotamus), aquatic marsupials (water opossum), aquatic rodents (beaver, water vole), and aquatic reptiles (crocodile, sea snake).

It is worth noting that all three subclasses of mammals (monotremes, marsupials, placentals) have thrown up secondary aquatic forms. The first subclass, the monotremes, are primitive mammals of which only two types remain extant. One of these, the duckbilled platypus, is an accomplished swimmer and diver with webbed feet.

Most surviving mammalian orders include species that took to the water and then evolved specific adaptations for aquatic life. One of the few orders that is generally believed to include no such species is the Primates—to which man belongs.

The aquatic theory postulates that one primate did follow that well-trodden path. This primate was the ape that was immediately ancestral to man.

During the period of the gap in human fossil history large areas of the northern half of the African continent were engulfed by the sea, apart from a few upland areas that formed islands. Later the water receded, the shallow marine areas were drained, and the islands once again became part of the mainland.

The aquatic theory envisages that during this period one group of apes embarked on a distinct path of evolution by adapting to an aquatic environment—just as other species had done earlier. Later, when the waters receded and new ecological opportunities opened up, they returned to their former terrestrial lifestyle. But they brought with them a package of inbuilt aquatic adaptations, which they still demonstrably retain. This has dramatically affected the course of their evolution ever since, and it accounts for most of the major differences between *Homo sapiens* and all other apes.

The theory suggests that man did not lose his hair because he became an overheated hunter, nor simply because the fetus of an ape is hairless at an early stage in its development. He lost it for the same reason as the whale and the dolphin and the manatee: because if any fairly large aquatic mammal needs to keep warm in water, it is better served by a layer of fat on the inside of its skin than by a layer of hair on the outside of it.

The three theories outlined above are all tenable. They are all derived by logical deduction from agreed data. Each has its own strengths and weaknesses. And in the absence of any complete fossil record, none of them is any more or less susceptible to "proof" than any of the others. The ensuing chapters will discuss what light can be shed by each of them on various features of human morphology and behavior.

II
Loss of Body Hair

1 The Naked Ape

Homo sapiens has been described as the naked ape, and this "nakedness" undoubtedly constitutes one of the most striking differences between the appearance of man and the apes.

Strictly speaking, man is not naked at all, being endowed with at least as many hair follicles per unit surface of skin as a chimpanzee. The only difference is that the hairs produced by man's follicles are for the most part shorter and finer and thus less conspicuous.

In view of this fact, some anthropologists consider it unnecessary to account for man's nakedness at all since it does not really exist. According to them, the difference here between man and apes is "only quantitative." This is a strange kind of reasoning. In that sense, almost all the differences between ourselves and the apes could be dismissed as unreal. Our skulls are only quantitatively larger, our stance only quantitatively more perpendicular. The use of such an argument suggests a wavering of faith in the conventional explanations and a disinclination to see the whole question reopened.

2 A Cooling Device

The savannah theory recognizes the need of an explanation, and its contention that the hair loss was a cooling device was for a long time the one most widely supported. Darwin considered it but was not convinced. He wrote in *The Descent of Man*: "The fact however that the other members of the order of Primates, to which man belongs, are well clothed with hair, generally thickest on the upper surface, is opposed to the supposition that man became naked through the action of the sun."

One of the chief weaknesses of this savannah explanation lies in the lack of any other example of a hunting mammal adopting this method of temperature control. The sun shines down as warmly on all the other inhabitants of the savannah. Hunting carnivores like the leopard and the cheetah run after their prey faster than any hunting primate can have done. Hyenas are as tireless over longer distances. Zebras and giraffes run as fast to escape being eaten. The camel, adapted to existence in extremes of heat and aridity, has retained its woolly coat, and it needs it at noonday just as much as in the chill of the desert night.

For all the evidence goes to show that hair is just as indispensable an insulation against excessive heat as it is against excessive cold. Its effectiveness lies in the fact that it traps a layer of temperate air next to the skin. W. P. Yapp, in *An Introduction to Animal Physiology,* comments: "The long wool of the merino sheep, which is adapted to hot climates, actually seems to prevent the animal from overheating, for when it is shaved off, at an air temperature of 30 degrees centigrade, the skin temperature rises by 3 degrees centigrade, and the rate of panting is doubled."

Using the same principle nomadic Arabs protect themselves from heatstroke by covering their bodies with robes and burnooses rather than by shedding all covering. African mammals possess this protective covering in natural form.

One of the few African mammals to have become hairless and thrived on it is the Somalian mole rat, a rare purblind subterranean mammal only a few inches long with white wrinkled skin. It has no need to fear the heat of the sun or the cooling of the wind because it spends its whole life burrowing under the sandy soil of East Africa. On the other hand it has no need to fear the chill of the desert night because the sand retains enough of the day's heat. It also need never come to the surface because it feeds on the roots of plants instead of their leaves. Its need of insulation to regulate its temperature is therefore minimal. However, this is an exceptional case, and whatever reasons *Homo sapiens* may have had for shedding his body hair, they are not likely to have been the same as those for the mole rat.

It used to be argued that the hunting ape's need for a cooling device was greater than that of any other carnivore because leopards and cheetahs had become adapted over many millions of years for running at high speeds and primates had not. A hunting primate was therefore faced with a unique problem and found a solution to it by shedding his hair.

One fallacy here is that on the savannah, a dangerous and competitive environment, the hunted have to run as fast as, or preferably faster than, the hunters. Other primates living on the savannah—vervet monkeys, geladas, patas monkeys, baboons—are all prey animals in the eyes of the larger carnivores. They have to be ready to run for their lives; they are not designed for speed, yet none of them displays even the first beginnings of a tendency to become naked for the sake of coolness.

A further weakness in the theory is the sexual dimorphism in respect of body hair in *Homo sapiens*. It was the male who was the hunter and allegedly became overheated in the chase—yet it was the slow-moving female, gathering food or awaiting the hunter's return, who became the most hairless.

3 Parasites

Another theory, dating back to Darwin's day, was attributed by him to a Mr. Belt, who claims that "within the Tropics it is an advantage to man to be destitute of hair, as he is thus enabled to free himself of the multitude of ticks and other parasites, with which he is often infested, and which sometimes cause ulceration." Darwin rejected this one, too.

His final comment was: "Whether this evil is of sufficient magnitude to have led to the denudation of his body through natural selection may be doubted, since none of the many quadrupeds inhabiting the Tropics have, as far as I know, acquired any specialized means of relief."

Belt's argument has sometimes been combined with the hunting one to suggest that man, having become a hunter, shortly afterward became a lair dweller. As a consequence of killing large game (more than could be eaten at one time), man would need somewhere to store it, and lairs when occupied over a period of time provide greater opportunities for parasites to thrive and breed.

Here again the weakness is that a great many animals occupy lairs, dens, setts, or burrows. Most of them become infested, yet they have not attempted to delouse themselves by turning into naked wolves and naked badgers. It is hard to believe, when a problem is common to so many species, that one alone has found a "successful" solution. Natural history, far more than the other kind, tends to repeat itself.

4 Disadvantages of Nakedness

Even if nakedness were an effective way of eliminating external parasites, this minor relief would surely have been far outweighed by the overwhelming disadvantages of hairlessness for a ground-dwelling ape. In the first place, body hair, besides being a highly efficient insulator against both cold and heat, serves as a protection against various kinds of lesions. A forest ape would certainly not have less need of such protection on moving to the savannah.

If man had been a pachyderm, like the elephant, the loss of hair would not have mattered, for the skin itself would have served as armor. But this was far from being the case. The ancestral ape-man must have been quite small when he left the trees. (His descendant, *Australopithecus*, millions of years later was not more than four feet tall.) And the smaller the animal, the thinner the skin. Max Kleiber has demonstrated that skin thickness increases slowly but progressively with body weight in mammals. His formula indicates the ape-man's skin would have been less than a third as thick as the elephant's.

If nakedness was disadvantageous to the hunting male, it must have been far more so to his mate and his offspring. A newborn ape can cling on with its hands to its mother's fur by the second day of life. The children of the naked ape would have been equipped with the instinct to cling but would have found no fur to cling to. The female, while food gathering or trying to run away in times of danger, would be immeasurably more hampered by an infant whose weight had to be supported by the strength of her own arm.

Jane Goodall has reported that one major cause of mortality among chimpanzee infants is injury incurred by "falling from the mother." The chances of this can only have been increased by the mother's skin becoming smooth and streamlined and, in rainy weather, slippery.

The disadvantages were at least sufficiently real to make it unlikely that nakedness would have evolved for such comparatively trivial reasons as the avoidance of fleas.

5 Sexual Attraction

This is the explanation that Darwin finally opted for. He wrote: "The view which seems to me the most probable is that man, or rather primarily woman, became divested of hair for ornamental purposes.... According to this belief, it is not surprising that man should differ so greatly in hairiness from all other Primates, for characteristics gained through sexual selection often differ to an extraordinary extent in closely related forms."

No one knew better than Darwin that sexual selection is one factor that sometimes operates to a point where it cannot be said to be conducive to the comfort or the convenience of the individual animal. For example, the tails of some species of birds of paradise have grown so long and elaborate that they hamper them considerably in flight and moving through branches. Yet as long as these adornments are helpful in attracting mates, they continue to be transmitted to future generations. Likewise, Darwin argued, nakedness might evolve on ornamental grounds, no matter how inconvenient it proves for day-to-day living.

On the other hand, there is one thing about this argument that fails to convince. "Characteristics gained through sexual selection" evolve by a process of exaggerating some feature that is *characteristic of the species*. For instance, all proboscis monkeys have longer noses than other species; in the adult male it burgeons to a magnificent size. Some species of moths have spots on their wings; in these species the males are irresistibly attracted to females experimentally endowed with larger-than-normal spots. But features that are *uncharacteristic* of the species are more likely to repel than attract. We have no reason to believe that a depilated chimpanzee or a bald cat would be considered ornamental by its own kind.

In all mechanisms for sexual attraction, natural selection acts not only on the signal (e.g. naked skin), but also on the receptor mechanism (the eye, the brain, and the aesthetic sense). "Beauty is in the eye of the beholder," and if we consider nakedness attractive in human beings, that preference has evolved as a response to the nakedness rather than as a trigger for it. Once hairlessness had begun to evolve for more utilitarian reasons, sexual selection might ultimately favor it—but it would not have initiated so bizarre and abnormal a development.

There is one last theory—or there was, for it is less often proposed nowadays—as to why a hunting primate on the savannah should have turned into a naked ape. It was a variant on the theme of sexual attraction, but it argued that the advantages of nakedness were tactile rather than visual. It was argued that a naked skin made sexual activity more pleasurable—"made sex sexier."

This sexual bonus was held to be more necessary to the hunting primate than it had ever been to the forest dwelling apes, and the reasons advanced for believing this were somewhat involved.

They hinged on the argument that when the prehuman primate took to hunting, the hunting was done by males and that they needed some special inducement to carry the kill back to the waiting females and share it out, rather than eat it on the spot. At the same time, pair bonding between one male and one female became adaptive because the young matured so slowly that it took two parents to share the burden of rearing them. Making sex sexier, it was argued, served the dual purpose of (1) ensuring that the males would willingly return to base and share the food so that the females and the offspring would not starve and (2) strengthening the pair bond.

Two fallacies are enshrined in this line of reasoning.

(1) It is now generally recognized that in any hunting/gathering economy the plant food gathered by the females provides the major and more dependable percentage of the total food supply. Food exchanges (meat versus plant food) would have provided quite adequate incentive for the hunters to return to base without the need of evolving unprecedented kinds of sexual allure.

(2) It has always been generally recognized that making sex more exciting does not necessarily favor monogamy.

6 The Naked Fetus

The neotenists do not feel the need to prove that nakedness was in itself either convenient, or attractive, or in any other way adaptive. They point out that there were powerful advantages to be gained by becoming paedomorphic (juvenilized)—that by this means man evolved to a higher form of life than the ape. He developed much more slowly, retaining into adult life such juvenile characteristics as a large skull, adaptability, curiosity, etc. And if, as part of this package, he also retained some features that were not so advantageous, they would nevertheless not be eliminated since the package as a whole was a good one. Man's nakedness, they would argue, was such a feature, and no further explanation of it is required.

This is ingenious and sounds reasonable. The fetus of an ape is, at an early stage of its development, quite naked.

But it is not quite as reasonable as it sounds. If the nakedness of the fetal ape were being retained into adulthood by a process of neoteny, one would expect the human body to keep this characteristic throughout its whole development from embryo to adult. But it does not do so. When the human fetus is in the sixth month of its development it becomes completely covered with a coat of fine hair known as lanugo. Normally this hair is shed long before birth, but occasionally a baby is born still wearing its woolly coat, only to lose it within the first few days after birth. More rarely still, in cases of an abornmality called hypertrichosis, this hairiness persists into later life and gives rise to rumors that in some remote tribal area "ape men" or "missing links" have been found.

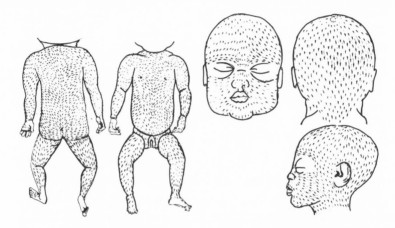

Fig. 3.   "*Left,* hair tracts of the dorsal and v ventral surfaces of the trunk of a human fetus. *Above,* hair tracts of the human face; *above right,* of the scalp; *lower right,* of the human head." From *Man's Place among the Mammals,* by Frederic Wood Jones (Edward Arnold & Co. Ltd.).
Reproduced from Sir Alister Hardy's article in the *New Scientist.*

   In fact, this coat of lanugo was one of the things that first suggested to Professor Alister Hardy the idea that man had at one time in his evolutionary history gone through an aquatic period. Professor Hardy had seen drawings by Professor Wood Jones of the tracts of hair on the unborn human child. The way they were arranged reminded him of the passage of water over a swimming body, and he speculated that the hair of an aquatic primate would have come to be arranged in that way before it finally disappeared (see Appendix 2). In any case, the very existence of this lanugo, regardless of the direction of the hair tracts, makes it logically impossible to accept neoteny as an explanation of this particular feature.

It is manifestly contrary to the whole theory of neoteny that a human fetus should first acquire the coat of hair once appropriate to an adult primate and then subsequently abandon it. It is as though the axolotl first acquired the lungs of a salamander and then discarded them; or as if it first shed its external gills and then immediately retreated back into juvenility by growing another pair. There are many examples of creatures retaining juvenile or fetal characteristics. There are no examples of creatures *regaining* such characteristics after they have been lost by the normal process of growth and maturation.

7 Hairless Aquatics

If we regard the ancestral primate as an aquatic ape, he ceases to be a mysterious zoological aberration evolving unique and inexplicable features of no use to himself and highly deleterious to his children. Put him among the aquatic mammals and he becomes a conformer, obeying the laws of evolution instead of running contrary to them.

Virtually all the hairless mammals in the world today are either aquatics, or wallowers, or show clear signs of having descended from aquatics or wallowers. (The exceptions are the mole rat—discussed earlier—and an artificially bred Mexican hairless dog.)

The longer an aquatic animal has been in the water, the more complete is the hair loss. Some cetaceans (Geoffroy's Dolphin and the La Plata Dolphin) still retain a few vestigial hairs around the mouth, but most are completely naked except for a few bristles where their whiskers used to be.

The hippopotamus spends all day in the river and only comes ashore at night. The rhinoceros lives on land but rolls and wallows in mud whenever it finds an opportunity. This habit is so ingrained that the so-called "white" rhinoceros is sometimes believed to have earned its name in this way. Its skin is no paler than that of the black rhinoceros, but it is most often seen plastered with a thick layer of mud that dries white in the sun.

The domestic pig is descended from the wild pig (*Sus scrofa*), a sparse-haired inhabitant of marshy woodland districts. This wild pig is an excellent swimmer, and its territorial range normally includes a favorite wallowing place. Another wild pig, the Babirusa, which also lives near rivers and swamps and swims readily, has a tough grayish hide almost as hairless as that of the hippopotamus.

The elephant bears so many and so diverse signs of being an ex-aquatic that he deserves a section to himself (Appendix 4).

The value of fur as an insulator lies in its ability to trap a layer of air next to the skin. When the fur becomes water-logged, its value as an insulator is destroyed. And not only does it lose its effectiveness. For a fully aquatic animal like the dolphin, fur would be a positive handicap, because it would detract from its streamlining and slow down its swimming. Alister Hardy points out (Appendix 2) that even in human swimmers the speed of their progress through water can be measurably affected by the amount of body hair *Homo sapiens* still retains.

The aquatic theory, then, attributes man's comparative hairlessness to the operation of convergent evolution. This factor is just as powerful as neoteny in the emergence of new adaptations. It is the process by which animals of quite separate origins grow to resemble one another when they adopt the same habitat or lifestyle. For example, mammals from three entirely different orders—the anteater, the pangolin, and the aardvark—have all, through becoming insect eaters, acquired long tapering snouts, powerful claws for digging, and extremely long sticky tongues. In the same way mammals from several entirely different orders have lost their body hair after changing to an aquatic habitat.

Not all aquatic mammals are hairless, but none of the factors making for hair retention would have applied to an aquatic ape. One major governing factor is size. The largest aquatics—such as the whale and the hippopotamus—are naked. The smaller ones—otter, beaver, platypus—have adapted to a watery habitat by changing the nature of their fur, rather than shedding it. In the case of one of the smallest, the African water shrew, the whole coat is so water resistant that when the animal dives he gives the appearance of being enclosed in a silvery bubble of air. An animal the size of *Australopithecus*—about four feet tall—would not have been small enough to follow that road. He would have been more or less identical in size with some of the smaller cetaceans (e.g. Commerson's Dolphin or the Finless Black Porpoise), all of which are totally naked.

Climate also probably exerts some influence here. Most of the pinnipeds (seals, sea lions) have retained their hair, possibly because they still return to land to breed and rear their young, often spending weeks ashore on cold and windswept Arctic beaches. An African ape would not have had to contend with this climatic hazard. For an animal of that size and in that situation moving into an aquatic environment, hair loss would have seemed likelier than not.

8 Naked Mothers

The care of infants in an aquatic environment must have posed some problems for an aquatic ape—as it must have for every aquatic mammal in the initial stages of adaptation to the water.

One effect of living in water is that the young are comparatively weightless. The manatee cub, for instance, a very slow-maturing infant with a birth weight (in air) of about 60 pounds, is cared for assiduously and with an appearance of effortlessness. In the days when manatees were much more numerous, voyagers to America often saw herds of many hundreds of manatees, with the females floating upright in the water holding the cubs in their flippers and watching the ships go by. Sometimes when the mother is feeding, the male manatee will cradle the infant—and then hand it back to her when she returns to suckle it.

The teats of the female manatee are located at the base of her flippers. Colin Bertram, in his manatee book *In Search of Mermaids*, says that during lactation the breasts are "large and shapely." Not all observers confirm this, so possibly it varies between individual animals or according to the stage of lactation. Steller, the discoverer of the now extinct

Steller's sea cow *(Hydrodamalis)*, also used the term "breasts" rather than "teats" when describing his discovery of the largest of the sirenians. He recorded that the breasts "are two, and pectoral," i.e. as in *Homo sapiens*.

This feature of the sirenians is the nearest parallel in the animal kingdom to the breasts of woman. Since it does not occur in other aquatics it cannot be claimed as an obvious instance of convergent evolution; but certainly nothing is found in the anatomy of monkeys and apes that even remotely resembles it.

However, we do not need to resort to the manatees for a description of aquatic mother/child behavior. *The National Geographic Magazine* in March 1975 printed a study of the behavior of aboriginals in Tierra del Fuego, which included the comment that "the women spend long periods in the water, with the children hanging on to their hair."

There would still have been, after all, something for the aquatic ape's child to clutch onto in the water when it needed a rest. This behavior pattern offers a possible explanation of the fact that scalp hair did not disappear together with body hair, and why in some races it grows longer than the hair on any part of any other primate. Scalp hair is not a feature of any other aquatic mammal, but since they are all believed to have descended from species with hooves or claws this is hardly surprising. Their infants would have been unable to cling onto the scalp hair even if it had been retained.

If the reason for retaining scalp hair was connected with nurturing, this might also explain why baldness is so much more common in men than in women. Since among Primates nurturing is carried out mainly by the females, female hormones would be expected to promote more permanent retention—as in fact they do.

It would, finally, offer an explanation of why women's scalp hair begins to grow thicker during pregnancy. In this case, individual hairs do not simply become longer or more lustrous; if that were the case, the hair thickening would doubtless be explained away as "for sexual attraction." This could hardly be the purpose here, since pregnancy is not a stage where additional sexual attraction could contribute in any way toward species survival.

What in fact happens is that at the onset of pregnancy thick hairs are produced in abundance, while the production of thin hairs is diminished. By the time the baby is born it has been provided with a stronger anchorage.

Meanwhile, the infant himself has evolved some adaptations to aquatic life, but these can be more conveniently discussed in later chapters.

III

Subcutaneous Fat

1 Fat Distribution: Aquatic and Terrestrial

Marine mammals that have shed their fur have replaced it by a method of insulation against heat loss that is more suitable for an aquatic existence, i.e. a layer of fat under the skin all over their bodies.

This subcutaneous fat serves a variety of purposes. It protects them against cold; it makes their bodies more buoyant; it stores energy; and it helps to give them a rounded, streamlined outline. It is so invaluable to them that even those species that have not become hairless have acquired it. Harbor seals, for example, have retained all their fur—and at the same time they have also acquired an additional fat layer. In Greenland Right Whales, this layer constitutes 40 percent of the body weight and in some porpoises as much as 50 percent.

Some marine biologists speculate that this feature may serve yet another function that remains to be discovered. They argue that sirenians, which are well endowed with fat, need it neither for insulation nor for food reserves—since they live in warm climates and the vegetation on which they feed is plentiful all around the year. Furthermore, for some deep-diving cetaceans buoyancy is no great advantage. Their ability to sink is as important to them as their ability to float. Whatever its functions, this layer of subcutaneous fat is one of the most ubiquitous characteristics of aquatic mammals and of wallowers such as the pig.

It is also one of the features that distinguishes *Homo sapiens* from all other Primates. It was this, even more than hairlessness, that first suggested to Alister Hardy the idea that our species might be ex-aquatic. He described in his *New Scientist* article (Appendix 2) how in 1927, shortly after returning from a two-year expedition studying marine life in the Antarctic, he came across the following passage in Professor Wood Jones' book, *Man's Place among the Mammals*:

"The peculiar relation of the skin to the underlying superficial fascia is a very real distinction, familiar to everyone who has repeatedly skinned both human subjects and any other members of the Primates. The bed of subcutaneous fat adherent to the skin, so conspicuous in Man, is possibly related to his apparent hair reduction; though it is difficult to see why, if no other factor is invoked, there should be such a basal difference between Man and the Chimpanzee."

This reminded Hardy of what happens when whalers or sealers cut into the dead bodies of their victims: they encounter in the first instance not muscle, but fat. The same feature in human bodies frequently complicates the work of surgeons and reputedly makes the conduct of postmortems in tropical countries a somewhat unpleasant exercise.

Terrestrial mammals also have fat in their bodies, but it is differently located—in the mesenteries (the membranes between the viscera) and around the kidneys, etc. In land animals, only a very small percentage of body fat is stored under the skin—just enough to keep it supple. These fat cells are clearly visible in cross-section under a microscope. As in the case of vestigial body hairs, the presence of the cells is sometimes used to argue that this further striking difference between man and other Primates is again "only quantitative" and therefore in some way unreal, thus calling for no explanation.

However, this fat is not only differently located in land animals—it also has fewer functions. Buoyancy is of no particular relevance to land animals; they have no special need to become streamlined; and in their furry coats they have a more efficient method of temperature control in air since heat loss can be regulated by raising or lowering the hairs relative to the skin surface.

In terrestrial mammals, then, the primary function of body fat is the storage of energy, and when a surplus of fat is acquired it is laid down in the sites referred to—the mesenteries, etc—and not under the skin. For example, a captive orangutan may become obese in old age. His excess fat will make him pot-bellied, but it will never, as with obesity in human beings, give him fat thighs and fat cheeks and fat fingers. Similarly, a hibernating bear or dormouse stores its food reserves in the abdomen. Some few terrestrial species with seasonal fluctuations in their food supply have evolved special arrangements for storage. Among reptiles there is a fat-tailed gecko; among mammals there are fat-tailed sheep and fat-tailed dormice. Even among primates this adaptation occurs in the mouse lemur and the fat-tailed dwarf lemur.

Only the aquatics and *Homo sapiens* dispose of a fat surplus by thickening the subcutaneous layer. Once again, if we regard man as a terrestrial mammal he is in this respect, as in others, somewhat freakish. But if we regard him as aquatic or ex-aquatic he is simply conforming to type.

Morever, this distinctly human trait is apparent at a very early age. Humans produce infants who weigh almost twice as much as those of the great apes (3.5 kg. as compared to 2 kg.: 7.7 lb. to 4.4 lb.). The face, neck, and limbs of a young chimpanzee present to our eyes a cadaverous appearance such as is only seen in human beings in conditions of very severe malnutrition or in extreme old age (see fig. 1). Viewed as an aid to buoyancy and heat insulation in water, the plumpness of the average human baby makes evolutionary sense. In any other terms it is very hard to understand. Whether in the trees or on the savannah the extra weight involved would take considerable toll of a primate mother; and the baby would certainly not need the fat for energy storage since at that age its food supply would be guaranteed for at least several months ahead.

2 The Savannah Explanation

The savannah theorists' answer to the problem is to postulate that (a) the hunting ape and his family first denuded themselves because the male was too hot, and (b) they subsequently (including the male) evolved a layer of subcutaneous fat because they were not warm enough.

It seems an eccentric procedure. The hair follicles were still there all over their bodies. If, for any reason—such as climate—the diminution of the hairs had turned out to be a mistake, it would have been relatively easy, in evolutionary terms, to lengthen them again. Primates can begin to adjust hair growth to temperature changes even within the lifetime of an individual. When monkeys from a warm climate are moved to a northern zoo like Moscow's, their hair grows perceptibly thicker.

However, the savannah explanation produces even further eccentricities. According to the argument put forward, once our hairless ancestral primate had taken to the African savannah and begun to endow himself with a coat of subcutaneous fat, he was now liable to become overheated again. And once again, an evolutionary development that was proving highly inconvenient was not halted or abandoned but countered with another new strategy. The hominid retained the fat layer and evolved a new way of cooling down, at need, by the use of sweat glands—between 2 and 3 million of them, far more than any other primate possesses. So now, when the temperature of *Homo sapiens* rises too high, the fluid secreted by these sweat glands covers the surface of his denuded but well-insulated skin with a layer of moisture that cools him down as it evaporates. Other primates have sweat glands in their skin, albeit fewer, and they react as ours do to emotional stresses such as fear. But they are not used as a cooling device. However hot a chimpanzee gets, its skin remains dry. *Homo sapiens,* among his other distinctions, is the perspiring primate.

3 Sweating

This development is often referred to as if it were a particularly elegant mechanism, so effective that perhaps it came first, and the hair loss was initiated solely for the purpose of making it work even better since fluid evaporates more quickly from a naked skin. Nearer the truth, surely, is William Montagna, Professor of Dermatology at the University of Oregon, when he argues that sweating as a means of temperature control represents "a major biological blunder," draining the body of enormous amounts of moisture and depleting the system of sodium and other essential elements.

Forest apes are such good moisture conservers that they rarely need to visit predator-haunted water holes. They get enough water from the fruit and vegetation they eat and from the rain that falls on the leaves. The savannah ape, though moving to a more arid habitat, apparently became profligate of these things.

Studies of man's water requirements in hot temperatures—which assumed military significance during the desert warfare in World War II—were carried out by a University of Rochester group under the direction of E. F. Adolph. They revealed that whereas a man requires 1 liter or less of water intake per day for the purposes of urine excretion, perspiration caused by walking naked in the sun at 100° F. can cause him to lose up to 28 liters per day. The daily salt loss in such cases can amount to 10-15 percent of the total quantity in the body; and even though the salt content of the sweat diminishes after a week or so of sweating, it never diminishes enough to prevent significant salt loss.

Montagna, then, is not exaggerating when he says that the sweat glands "are still an experiment of nature, not fully refined by the evolutionary process." Nature does sometimes bungle her experiments. But it would be untypical of her, to say the least, to encourage one originally rather nondescript little primate to embark on an inappropriate program of skin denudation, countered by an equally inappropriate subcutaneous insulation, countered again by a biochemically costly program of sweat-cooling—all to cope with a climate that every other savannah dweller had accepted without turning, or shedding, a hair.

4 An Ex-Aquatic Adaptation

The aquatic theory suggests that hairlessness and subcutaneous fat were both entirely appropriate to an aquatic animal—were both, indeed, normal for such an animal.

For an aquatically adapted mammal to return to terrestrial life is extremely rare and would probably only take place as a result of an exceptional set of circumstances, which are discussed in a later chapter.

The aquatic primate returning to land would certainly face a unique situation. The subcutaneous fat layer, by then well established, would create serious problems of temperature control for him. However, assuming that he still regarded water as an element that he could master and felt safe in, he would have stayed close to the rivers in any case.

In those circumstances, perspiration as a method of temperature control was a neat enough solution to his problem since the fluid loss could be easily made up.

5 Neoteny Inapplicable

The sequence of events proposed by the aquatic theory seems somewhat more credible than that proposed by the savannah one. In this connection neoteny cannot begin to explain the plumpness of a baby, since no ape ever had a plump fetus.

And neoteny cannot begin to explain sweating. Babies, far from retaining the habit of an earlier fetal stage of development, are born without the capacity to sweat. They are incapable of regulating their temperature either by sweating or by shivering until they are several weeks old.

IV
Tears

1 The Weeping Primate

All mammals have tear glands, whether they weep or not. The function of these glands is to secrete a very small but constant flow of saline and mildly antiseptic fluid, which coats the surface of the eyeball with moisture, thus acting both as a protection and a lubricant.

In a very few mammals the supply of this fluid may be considerably augmented at moments of emotional agitation so that it overflows from the eyes in the phenomenon known as weeping.

Man is the only weeping primate. Indeed, with one exception (see Appendix 4), he seems to be the only terrestrial mammal of any kind that sheds tears.

The chimpanzee, for example, remains permanently dry-eyed—but not because it is phlegmatic. The *Larousse Encyclopedia of Animal Life* comments: "The range of its emotional life is considerable, and it is able to express surprise, interest, disgust, fear, anger, joy, sadness, and even despair, the latter being evinced by fits of sobbing. Only tears are lacking."

In nineteenth-century discussions about our origins, this peculiar human attribute gave rise to a good deal of puzzlement and speculation, but during recent decades it has seldom been mentioned by professional anthropologists.

This is not because the problem has been solved. On the contrary. As Professor Peter Medawar has pointed out: "Scientists tend not to ask themselves questions until they can see the rudiments of an answer in their minds." And Victorian scientists, like one Victorian poet, had to content themselves with the admission: "Tears, idle tears, I know not what they mean."

2 Sea Birds

For a long time a cognate mystery about sea birds remained similarly unsolved. The question in the case of the sea birds concerned not tear glands but two glands in the bird's head which drain into the internal nares (nostrils) and are known to anatomists as "nasal glands." In sea birds the glands are much bigger and more highly developed and have a richer arterial blood supply than in birds that live inland. Even within a single genus such as the gull, those species that spend the highest proportion of their lives at sea have the largest nasal glands.

In 1956 Knut Schmidt-Nielsen carried out an investigation into the salt and water balance of the double-crested cormorant. As part of his investigation, he fed cormorants by stomach tube with a quantity of sea water amounting to approximately 6 percent of the birds' body weight. The object was to find out how a species that lived by catching and eating salt-water fish dealt with the consequences of the incidental ingestion of sea water that must inevitably occur.

When the sea water was administered, more salt was excreted in the bird's urine. This had been expected. What had not been expected was the secretion by the nasal glands of a clear waterlike liquid that ran from the nasal openings and down the beak to accumulate at the tip, from which the drips were shaken off by sudden jerks of the head. On analysis, the drops thus secreted turned out to be an almost pure solution of salt. These were tears far brinier than our own.

The same experiment produced the same results in the cormorant, the herring gull, the Humbolt penguin, and the albatross. Microscopic examination of the gland's anatomy in the pelican, the eider duck, the gull, and the petrel indicated a similar function for the nasal glands.

Further investigations proved that sea birds were not alone in this. If the marine loggerhead turtle swallows salt water, salt tears flow from its eyes. The same thing happens to the smaller *Malacolemys* terrapin.

There had long been an unresolved argument between those who claimed that a crocodile really does weep crocodile tears and others who knew crocodiles intimately and scoffed at this legend as an old wives' tale. It was finally settled by the discovery that both were telling the truth. Fresh-water crocodiles do not shed tears; marine crocodiles do. Land lizards do not weep, but the marine iguana of the Galapagos does.

It may be argued that it is begging the question to use words like "weeping" and "tears" for this phenomenon, because it is different in several ways from human weeping.

3 Salt Glands and Tear Glands

One obvious and genuine distinction is that the fluid shed by marine birds and reptiles has a higher concentration of salt than human tears. Our own tears are only moderately saline. They are not triggered off by drinking too much sea water. Even if they were, they would not help to redress the salt balance in our bodies because they are not sufficiently concentrated.

However, it would be misleading to give the impression that the tears of birds and reptiles are triggered off only by a physical stimulus (excess salt) and ours only by a psychological one (emotion). This distinction is not a real one because weeping can be induced by both kinds of stimulus in both species. Human tears flow freely in response to certain chemical irritants such as the vapor exuded by a cut raw onion, as well as in response to emotion. And the tears of marine birds flow freely in emotional situations as well as following a disturbance of the salt balance.

Homer Smith records of the albatross: "Nasal dripping was observed to occur when the birds had been fighting with each other, during their ritual dancing, or even during the excitement of feeding time."

A second distinction is that the birds' tears are excreted via the nostrils, whereas our own—like the crocodile's and the turtle's—are normally excreted from the eyes.

Our own tears, in fact, have a choice of outlets. The normal flow of fluid secreted to bathe our eyes in nonemotional moments does not emerge from the eyes at all. It makes its way out through a small duct in the inner corner of the eye called a nasolacrimal duct, which directs it into the nasal cavity and out via the nose. The quantities involved are so small that we are not conscious of this happening. The fluid helps to keep the inner surfaces of the nasal cavity moist to facilitate scent perception. It is ultimately exhaled as vapor.

A blockage in the nasolacrimal duct may cause the affected eye to water. On the other hand, a person afflicted by a blocked tear duct and simultaneously moved to emotion may well find himself applying a handkerchief in turn to his right eye and his left nostril. Also, when tears flow freely, one exit may not be enough so that wiping the eyes is frequently followed by blowing the nose. In cultures where men have been conditioned not to weep, the nose blowing may be a substitute for eye wiping and a sign that the inhibition is strong enough to rechannel the tears though not strong enough to prevent their production.

The actual point of emergence, then, seems fairly fortuitous even within our own species. The different points of exit for a marine bird and a marine crocodile are by no means proof that their tears are dissimilar in origin and function.

The third distinction between our weeping and the birds' nasal dripping is that the fluid is secreted by a different gland. In the case of birds and reptiles the saline fluid comes from the salt glands, whereas in ourselves it comes from the tear glands. But when we move from birds and reptiles to mammals, this distinction ceases to apply.

4 The Weeping Mammals

The tears produced in very small amounts by all mammals for the purpose of lubrication are secreted, in most species, by the tear glands. In some species (e.g. ruminants, elephants, and whales) they are secreted by the Harderian glands, which, however, serve the same purpose.

A study of the mammals who actually *shed* tears (as opposed to secreting them) supports the hypothesis that there is a strong connection between weeping and a marine habitat and also that among mammals emotional stress is the chief stimulus to the shedding of tears.

R. M. Lockley, in his book *Grey Seal, Common Seal,* commented that their tears "flow copiously, as in man, when the seal is alarmed, frightened, or otherwise agitated."

This is a significant observation. Seals have no nasal ducts, and their tears are sometimes dismissed as being attributable solely to this fact. But the amount of moisture bedewing their eyes from this cause would be no more "copious" than the amount bedewing our own nasal passages, i.e. negligible.

If you examine a hundred photographs of seals in the wild you are not likely to find more than one in thirty that shows traces of tears—even though in all cases the photographer must have been near enough to take their picture and thus conceivably to disturb them. You may watch well-fed and well-adjusted zoo seals all day long and never catch them weeping. But if a seal is distressed, or a mother seal bereaved, then the tears, as Lockley observed, flow thick and fast (see plate 5).

Other sea mammals have acquired the same habit. Steller, who specialized in studying marine mammals, wrote of the sea otter: "I have sometimes deprived females of their young on purpose, sparing the lives of the mothers, and they would weep over their affliction just like human beings."

On the other hand, terrestrial mammals (such as dogs and horses) do not express their emotions by weeping. No matter how deep their grief and how graphically they express it in other ways, they simply do not have this outlet for their emotions. The so-called "weeping" Capuchin monkey gets his name, like the "spectacled" monkey, from the pattern of the color on his fur. And the suborbital pits of certain male deer, once known as "tear sacs," are purely sexual in function. These exude a very powerfully smelling gummy secretion, not a fluid—and in castrated males this process does not develop at all.

5 Excretory Function

"Outlet for the emotions" is a phrase that may sound more sentimental than scientific. But an American psychiatrist, Dr. William Frey, together with his colleague Dr. Vincent Tuason, both from St. Paul-Ramsey Medical Center in Minnesota, have made a study comparing tears produced by distress with those produced simply by chopping onions. Their results have confirmed earlier findings that there are biochemical differences between the two kind of tears. Tears of emotion contain different proteins.

Dr. Frey argues that evolution seldom produces a purposeless function and that tears, like urine, are products of the exocrine system used to carry away wastes—presumably the chemicals produced in the body by stress. It is possible that this mechanism, developed in some marine species primarily for salt excretion, acquired the secondary function of eliminating other waste products secreted during stress and has been retained in our species to fulfill this function.

Whatever the purpose, we can say of human tears as of the hairlessness and the subcutaneous fat: man is the only primate possessing this characteristic. If we view him as a land animal, his possession of it is unique and inexplicable. If we view him as an ex-aquatic, he is conforming to the general pattern.

In this connection the aquatic theory is the only one that even attempts to explain this striking difference between ourselves and all other primates. The savannah theory has nothing to offer. No other savannah dweller sheds tears.

Neoteny has nothing to offer, either. A human infant does not bring weeping with it as an inheritance from its prenatal existence. Indeed, for the first few weeks after its birth it cannot weep at all. Like the chimpanzee, it manifests distress in a variety of other ways, such as yelling, wailing and grimacing. As with the chimpanzee, "only tears are lacking."

V
Bipedalism

1 The Perpendicular Ape

Man is the only mammal whose normal method of locomotion is to walk on two legs. A pattern of mammal behavior that emerges only once in the whole history of life on earth takes a great deal of explaining.

2 Alleged Advantages

Some of the more popular expositions of the savannah theory have seemed to imply that for the savannah ape to become bipedal was the most natural thing in the world. He took to walking upright, it is alleged, because it is faster and easier than walking on four legs and because it became necessary once he began to use tools and weapons.

For human beings today it is faster and easier to progress on two legs than on four because we have been progressively adapted for it over millions of years. What we have to explain is why bipedalism was resorted to in the first instance by a quadruped. The only examples of quadrupeds that can move faster on two legs than on four are those that have large, heavy counterbalancing tails, like kangaroos and wallabies.

Normally they progress by hopping rather than running, though there is a small long-tailed lizard called the Texas boomer that will sprint for short distances on two legs.

However, no primate possesses a tail that, in its relation to body weight, can compare with that of the kangaroo or the Texas boomer. Apes have no tails at all. Chimpanzees and gorillas, while perfectly capable of standing and moving on two legs, can proceed with considerably more speed and facility on four and always use that method when escaping from danger or covering long distances.

To establish whether or not bipedalism was actually easier, C. R. Taylor in 1970 embarked on a series of experiments designed to measure the relative energy costs of running on four legs and on two.

His first results seemed to prove that bipedalism was very costly in energy terms. He discovered that a man uses about twice as much energy to move 1 gram of body weight 1 kilometer as does a quadruped of the same weight. His study of another bipedal creature, the ostrichlike Rhea, appeared to confirm that running on two legs consumes twice as much energy as running on four.

However, this assumption had to be modified as a result of his later experiments, in which he measured the relative energy consumption of bipedal and quadrupedal locomotion *in the same animals.* He trained two chimpanzees and two Capuchin monkeys to run on a treadmill either on four legs or on two and measured their energy consumption in both cases and at various speeds. He discovered that they used neither more nor less energy by moving on two legs rather than four. The only difference was that on four legs they could run a lot faster. He advised that in future arguments about bipedalism the energy-consumption factor could be safely ignored.

According to the savannah theory, then, the ape on taking to the plains and becoming a hunter changed to a mode of progression that, though neither more nor less energy-consuming, was slower. It was also, and still is, more unstable. Millions of years of practice and physiological modifications have made bipedalism in *Homo sapiens* a rapid and efficient mode of locomotion; but not even those millions of years have eradicated the instability. The chances of tripping and falling, and the damage likely to be incurred by such accidents, are all much greater in our perpendicular species than in quadrupeds of the standard design with one leg at each corner.

There are other disadvantages too. These may be less visible, but over the course of a lifetime they are even more debilitating. The pattern of the muscles of the human body, and the arrangement of its internal organs, originally evolved to suit the requirements of a quadruped. Since we began to walk with our spines vertical instead of parallel to the ground, this arrangement does not function nearly as well. A long dreary catalogue of physical disorders—muscular strains, prolapses, hernias, backaches, and problems of the legs and feet—reflects part of the price we still pay for walking upright.

Thus, for the human anthropoid, bipedalism would have been slower, more precarious, and no less strenuous. It would also have imposed unaccustomed physical strains on a body designed to be quadrupedal. The motivation to adopt it would have had to be a very powerful one to outweigh such disadvantages.

Two motives are commonly suggested.

One is the need for the savannah-dwelling ape to leave his hands free for carrying tools and weapons. To evaluate this suggestion we need reliable fossil evidence indicating when bipedal locomotion began to emerge and also when tool using began to emerge. The latter is particularly difficult to deduce from the fossil record since the first things used as tools would have been naturally-occurring objects such as sticks and stones. And even when toolmaking had begun, only the durable kinds of tool would be fossilized.

But it may be observed that any ape, such as a chimpanzee, would have no problem in carrying around small portable objects without altering his mode of locomotion in order to do so. He could run off with a stick or a stone as easily as he runs off with a banana, holding it in one hand and running on the other three limbs—still much faster than he could on two. In fact, since chimps and gorillas have an unusual method of locomotion known as knuckle walking, which leaves the fingers largely free, they are able to carry objects in their hands while proceeding on all fours.

The actual use of tools, whether in mammals or men, is not often combined with locomotion. The sea otter lies on his back in the sea to open the sea urchin on his chest; the chimp squats down to poke her stick into crevices to tease out insects (this particular piece of tool using seems to be a female speciality); and a man will sit or stand to wield his hammer and chisel, knife or saw. Activities of this kind would never have led to bipedalism, since an ape's hands are already as free as our own for manipulating tools whenever he is stationary. The hunting ape would not find a need to *combine* tool using and locomotion until he came to throw some kind of missile at an animal he was chasing; and even then a running ape, carrying a stone or a spear, would have to stand still in order to hurl it. If he desired speed between the throws his best option would still have been to drop onto all fours.

Another suggestion is that on moving to the savannah the ape stood upright in order to see further. On the open plain it is possible to see long distances when searching for a glimpse of likely prey or predators. Here height is certainly an advantage. Prairie dogs on the American plains will sit upright, like rabbits, to scan the horizon. Vervet monkeys on the African savannah will stand quite upright, and even take a couple of stiff-legged sideways hops, while maintaining that posture.

But having seen what it wants to see, the vervet (like the gopher and the rabbit) drops onto four legs and uses them all to run with, if running seems advisable. Except for man, no mammal, carnivore or herbivore, on the savannah or off it, has chosen to walk and run habitually with its spine at right angles to the earth.

The most recent hypothesis for explaining bipedality is that of Owen C. Lovejoy (1981), who suggests that it was brought about by a supposed change in reproductive strategy, causing the male to range far afield and carry back "significant amounts of food" to females and young. It is not made clear what type of food the animal is supposed to have carried. If it was meat, a primate would have found it less tiring to follow his usual practice by proceeding on three limbs and dragging, for example, a haunch of zebra along with his spare hand. If it was vegetable food— roots, nuts, berries, or leaves—it envisages the primate making fairly extensive journeys to separate foraging areas in order to return home bipedally carrying two handfuls of food to supplement the supplies gathered by the female and her young.

On the face of it, it sounds a rather inefficient system, and it is hard to imagine the precise ecological circumstances that would have led to its being adopted.

3 Upright in the Water

Alister Hardy's original paper (Appendix 2) suggested that the first impulse toward bipedalism came when the ancestral primate waded into the sea. It would not have been able to advance very far into the water on four legs and still keep its head above water: the natural reaction would have been to stand up and proceed on two.

Possible reasons why the primate may have chosen, or been forced, to resort to an aquatic life will be discussed in detail in a later chapter. But there is certainly support for Hardy's contention in what we know of the behavior of ground dwelling primates that live near the shore line. Crab eating macaques, for example, have overcome the fear of water that is the characteristic of so many primates. They will wade into the water up to their waists hunting for sea food.

Another species of Japanese macaques being studied by American scientists on an off shore island acquired the habit of carrying food supplied to them (potatoes) down to the water to wash the dirt off. They stood up in the water to wash the food, and they walked on their hind legs while taking it down to the water.

Another aquatic example is the beaver, a rodent. When rodents move their young, they are in the habit of transporting them by mouth. But the beaver uses an upright bipedal shuffle, carrying the young on its front legs; and the male often uses the same method for bringing the building materials to erect or repair his dam.

The next point to consider is what would happen when the aquatic ape ventured out of his depth. If he was by then sufficiently accustomed to the water not to be alarmed, he would soon find that a moderate expenditure of effort in "treading water" would be enough to keep him vertical and his head above water.

Fig. 4. The male uses the same method for carrying building materials.

Jan Wind, in his paper *Human Drowning: Phylogenetic Origin,* (1976) notes that: "In animals, as in lifeless objects, the center of gravity during immersion tends to become fixed perpendicularly below the center of the upward forces. In man this results in a vertical or semivertical position, the former center being located in the upper abdomen, the latter usually in the lower thorax." This fact has been confirmed in papers by Donskoi, 1961; Slijper, 1962; Scott, 1963; and Cooper, 1968.

The adoption of a vertical posture in the water is favored by a wide variety of sea mammals when they are near the shore and something

Fig. 5. Vertical in the water is a favorite posture.

rouses their curiosity. Seals will float in this position, staring fixedly, for up to half an hour at a time. Sea otters do it to gaze at a passing boat. Dugongs and manatees do it, cradling their offspring in their flippers—supposedly one reason why sailors called them mermaids. And aquarium dolphins do it while waiting for food or instructions from their trainers.

4 Horizontal in the Water

Now, when a seal or a sea otter, or a man, gets tired of gazing and swims away, the position changes. Instead of being vertical in the water they are horizontal in the water. But one thing does not change, and that is the relationship between their spine and their hind limbs. While they are in the water, *whether horizontal or vertical,* the spine and the hind limbs are aligned in one straight line quite different from the 90-degree angle of a land dwelling quadruped.

Since a seal spends most of its life with its spine and hind limbs thus aligned, it is not surprising that something has happened to its pelvis. In a seal the pelvic girdle is more nearly parallel to the vertebral column than it is in terrestrial mammals. A similar shift can be observed in the pelvis of *Homo sapiens,* and in his case it is described as an adaptation to bipedalism. We may assume that it happened also to whales and dolphins, but this would be harder to prove since in their case only the vestiges of the pelvic bones remain.

If this modification of the skeleton of man's ancestors—a preadaptation to bipedalism—took place in the water, it would not be unique: it would be a common and natural development. It would not be a precarious one, either: being vertical in the water does not lead to instability and falling down.

If after a few million years of aquatic life—probably wading at first, and subsequently floating and swimming—the primate returned to the land, he would be already endowed with some physiological adaptations making the erect posture easier to adopt and maintain.

There is a parallel to this. One other creature habitually walks in the same posture as man, with its spine at right angles to the earth and its

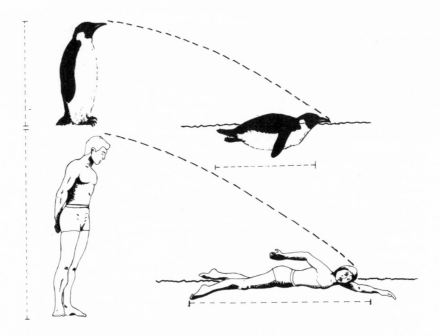

Fig. 6.   The penguin's perpendicular stance is as unlike that of any other
          bird as ours is unlike that of any other mammal.

skull in a direct vertical line with its feet. That creature is the penguin. Its
perpendicular stance is unlike that of any other bird, just as ours is unlike
that of any other mammal. The reason could well be the same. The
penguin's quasi-human gait, which so often causes us to laugh, could be
another example of convergent evolution.

5 Balance

Walking erect on land is a very difficult procedure. We have become so
adept at it that we underestimate the difficulties; but watching a young

child struggling to acquire the art reminds us how easy it is for the center of gravity to go astray and bring the biped tumbling down.

Some further preadaptations may have been necessary, or at least desirable, to make bipedalism a practical possibility: a more finely attuned sense of balance and a more flexible spine than possessed by our nearest relatives the gorillas and chimpanzees.

Both of these endowments are extremely common among aquatic animals, perhaps because they live in a three-dimensional medium. In such conditions a sense of the vertical is imperative, and under water it must be extra acute because there are no clues from the senses of sight and touch (a view of the horizon, the feel of the ground underfoot) to indicate to the animal which way is up.

The most flexible spine in the mammal world is possessed by the elephant seal, which can bend over backward with its spine not only forming a U-shape but something spectacularly like a V-shape. Olympic

Fig. 7.   A more finely attuned sense of balance and a more flexible spine.

gymnasts performing slow back-somersaults on the bars come close to rivaling its suppleness, but for any other of the great apes such feats would be quite impossible.

In many sea mammals the sense of balance is so keen that they take delight in exercising it. Dolphins in aquariums play with beach balls; the wild sea otters balance bits of kelp and driftwood for the fun of it. A circus sea lion can quite easily be trained to hoist itself up on one side of a flight of steps and down the other side, while balancing a ball on its nose—and never be in the slightest danger of letting the ball drop. A man can train himself to perform a similar feat, although it is harder for him. No other terrestrial mammal can do it at all.

6 Savannah and Neoteny Theories

An aquatic ape returning to land would have possessed a pelvic structure modified in the direction of bipedalism; his more flexible spine and enhanced sense of balance would make the erect posture much easier for him to sustain. He would be a very different animal from a forest ape coming down from the trees, and after so long an aquatic interval he would not necessarily find quadrupedal locomotion any easier or quicker or more natural than bipedal.

Under the circumstances the initial barriers that have deterred all other land mammals from adopting bipedalism would be considerably diminished. The advantages outlined in the savannah theory (greater range of vision, tool carrying, etc.) would no longer be outweighed by the difficulties. Once aquatic preadaptation had made bipedalism a tenable option, a subsequent existence on the savannah as a hunter would provide every incentive for adopting and perfecting it.

The neoteny theory provides some clues as to how the transition to bipedalism could have been fairly rapidly achieved. An erect posture requires that the head should bear a different relationship to the spinal column, and the foramen magnum (the opening through which the spinal cord passes into the skull) needs to be more centrally placed than is the case with quadrupeds.

In the early stages of the prenatal development of the skull of a fetal ape, the foramen magnum is in fact more centrally placed, and migrates backward as growth proceeds. Here as in other instances, neoteny explains the "how" but not the "why" of any evolutionary development. Many other species possess the same genetic potential for carrying their heads at a human-type angle. An environmental explanation is still needed as to why we are the only primate in which this potential has been exploited.

VI
Copulation

1 Face to Face

The most usual mode of copulation in all human cultures is face to face. The scientific term is ventro-ventral—literally, belly to belly. The species is physiologically adapted for this approach. The female sexual canal is tilted forward to accommodate it; this canal lies at a different angle in other anthropoids.

2 The Savannah Explanation

The savannah theorists explain this anomaly along the familiar lines of making sex sexier for the hunter when he returned from the hunt, and the need of cementing the pair bond. The argument here is that an approach that entails—or, at least, permits—gazing into one another's eyes is less impersonal than the rear entry method practiced by all other mammalian savannah dwellers and therefore more conducive to monogamy.

It is a plausible hypothesis. Its weakness lies in the fact that a great many different primate species practice pair bonding, but no other species has chosen this particular method of cementing it. The most successful and faithful pair bonders among the apes are not ourselves but the gibbons. They live in small family groups, and they mate for life. They manage to cement the bond very effectively without face to face sex or elaborate ornamentation for sexual attraction, or shedding body hair to enhance tactile stimulation. The relationship between the sexes is egalitarian; the females are rather aggressive; and the males are noted for their "very low sex drive." In fact, in all monogamous primates (siamangs, marmosets and tamarins, owl monkeys, indri) sexual behavior seems to be somewhat reduced rather than being intensified.

As for the hunter being lured home to base, here again there are numerous animal species where the male hands over food or cooperates in nurturing the young, and in none of these are special inducements offered to bribe him to do his duty. Wherever it is necessary for survival of the offspring, the instinct for parenting is implanted in the male as directly and unconditionally as in the female. Several of these savannah-based arguments appear to be based on present-day cultural conventions, rather than on anything we know of primate behavior in general.

3 The Aquatic Explanation

The aquatic explanation is simply stated. Once again it consists of the observation that *Homo sapiens,* considered as a terrestrial mammal, exhibits a unique and bizarre departure from the norm in his mode of copulation. Considered as an aquatic or ex-aquatic primate, he is merely conforming to the standard practice. The great majority of all marine mammal species behave in this way: they copulate face to face, and the females have ventrally directed sexual canals. It is a direct consequence of the development outlined in the previous chapter: the realignment of the spine and the hind limbs in the same straight line.

Since man's most typical method of mating was for many years confidently described as "unique," a few quotations are supplied to establish that this is not the case.

Victor Schaffer, in *The Year of the Whale,* describes the mating of the largest living mammals. "Hour after hour the pair swim side by side, keeping in touch by flippers and flukes, or simply rubbing sides. . . . Presently the male moves to a position above the female, gently stroking her back. . . . The cow turns responsively upside-down, and the bull swims across her inflamed belly. . . . At last the pair rise high from the sea, black snouts against the sky, belly to belly, flippers touching, water draining from the warm, clean flanks. They copulate in seconds, then fall heavily into the sea with a resounding splash."

Dolphins are nowadays so common in seaquaria that their mating behavior has frequently been described. It is ventro-ventral and, according to R. M. Martin in *Mammals of the Sea,* is "a wonderfully refined and sensitive affair with what seems like real love transmitted between the two participating individuals."

Colin Bertram, in *Search of Mermaids,* quotes an eyewitness account of the mating of manatees. "The manatees were disporting themselves in the river toward the left bank and in a school of 14 or 16; they gave the impression of fighting among themselves. Later they moved into the shallow and worked themselves up the bank into six inches of water. One pair was completely out of the water. They mated lying on their sides." In the case of both manatees and dolphins the position of the female sexual organs makes any approach other than ventro-ventral totally impracticable.

The mating of beavers is not often seen, but it has been observed on Russian breeding farms. Lars Wilsson gives a description of it. "The scent of the female in heat is presumably enough to make the male sufficiently stimulated, and when she goes into the water in a particular way he follows, and mating takes place stomach to stomach, the animals swimming slowly forward."

Pinnipeds (seals and sea lions) are an exception, at least when they mate on shore. The male then approaches from behind and mounts the female in the normal fashion of terrestrial quadrupeds.

In the water it may be different, for David Maxwell gives this description of the behavior of seals in the Shetland Islands. Pairs of seals "roll and twist in the water, snarling and snapping at each other. . . . This play is thought to be equivalent of courting, as copulation appears to occur afterward, the cow rolling on her back with the male on top gripping her with his flippers."

Steller's sea cow *(Hydrodamalis)* was a large sirenian once plentiful in Arctic waters, and before it died out Steller wrote this eyewitness account of its behavior.

"In the Spring they mate like human beings and especially towards the evening, if the sea is calm. Before they come together many love games take place. The female swims slowly to and fro, the male following. He deceives the female by many twists and turns and devious courses until finally she herself becomes bored and gets tired and forced to lie on her back, whereupon the male comes raging toward her to satisfy his ardor and both embrace each other."

All these sea mammals are descended from land dwelling quadrupeds that presumably mated from the rear in the normal mammalian way.

The change of approach is not always accomplished without some inconvenience to one or both partners. That suffered by the female sea otter, for instance, is visible and dramatic. The otter's nearest terrestrial relatives are the stoats and weasels. Presumably, before returning to the water, the otter shared their mating behavior. This involves the male sinking his teeth firmly into the thick fur at the back of the female's neck and holding on throughout an act of copulation, which may last for 15 minutes. Since they became aquatic, sea otters have made the usual transition to ventro-ventral copulation, but the male's instinct to hold on with its teeth is too deep rooted to be abandoned. The result is that in any group of sea otters during the mating season the females already mated can easily be recognized: their noses are covered with blood.

In our own species much of the sexual malfunctioning, especially in females (usually attributed to some kind of psychological disturbance), may well have its roots in the simple physical fact that our sex organs, like our spinal columns and the arrangement of our muscles, were basically designed for the convenience of quadrupeds. The once efficient provision for mutual gratification has not functioned quite so smoothly since the changeover.

4 The Neoteny Explanation

Professor Louis Bolk, a professor of human anatomy at Amsterdam, was one of the earliest propounders of a version of the neoteny theory. He summed up his beliefs by saying: "If I wished to express the basic principle of my ideas in a somewhat strongly worded sentence, I would say that man, in his bodily development, is a primate fetus that has become sexually mature."

In one of his books he listed some of the human features attributed by him to the retention of fetal characteristics. The list includes our "flat faced" physiognomy, reduction or lack of body hair, high relative brain weight, structure of the hand and foot, etc. (The complete list can be found in Appendix 3.) The items of relevance here are "the form of the pelvis" and "the ventrally directed position of the sexual canal in women."

Not every biologist is convinced of the validity of these particular items. Bolk does not specify what he means by the form of the pelvis, nor at what stage of its development a fetal or juvenile ape has a ventrally directed sexual canal. Stephen Jay Gould, a strong supporter of the neoteny theory in general, comments that the reference to the pelvis is, ". . . I confess, vague and confusing. But I have translated Bolk's words literally."

And Dr. R. D. Martin comments: "Since it is a primitive land vertebrate characteristic to have a cloaca, with the reproductive, urinary, and digestive tracts all running back to a common backward-facing outlet, I find it hard to believe that a ventrally-oriented sexual canal in the female could be an early embryonic feature."

However, now that faith in the savannah explanation is beginning to waver, Bolk's observations are being cited as evidence that neoteny can supply an adequate explanation of man's mode of love-making. The explanation runs as follows:

(1) The ventrally directed female sexual canal is a characteristic of a fetal ape.

(2) Man evolved from apes by a process of paedomorphosis because certain juvenile features were more favorable to his survival than the features of a mature anthropoid.

(3) Together with these favorable characteristics, man inherited, as part

of the neotenic package, some traits that were not necessarily of any intrinsic value to him.

(4) One of these was the ventrally directed female sexual canal.

(5) As a result of acquiring this modification in the female, *Homo sapiens* was forced to adopt a ventro-ventral mode of copulation.

(6) No other explanation is required. The aquatic theory is superfluous.

That is neat, but it leaves one important question outstanding.

If neotenists claim that ventro-ventralism in human beings came about because of a characteristic of the primate fetus, they still need to explain why the same development took place in the whale, the dolphin, the sea otter, the dugong, *et al.*—not one of which has a primate anywhere in its ancestry.

VII
Swimming and Diving

1 The Primate Fear of Water

When the aquatic theory was originally propounded, many people rejected it instinctively rather than on rational grounds. The reasons they gave were that (a) primates in general have greater fear of water than most other mammalian orders; (b) many human beings nowadays are unable to swim; and (c) it was believed at the time that the water would be a particularly perilous environment for young babies.

The first of these is certainly true, though a few primate species (talapoins, crab-eating macaques, and particularly proboscis monkeys) do not share this fear of water. The great apes have a powerful dislike of it so that a moat around their enclosure in a zoo is usually as effective as a fence or a cage. True, there was one gorilla in an African zoo who learned to love his moat and use it as a private swimming pool, but he was an exception.

The primate fear of water is easily explained. In the case of a ground-dwelling animal, such as a deer or a hyena or a lion, the first encounter with a river or lake is likely to be from the bank, giving an opportunity for tentative and leisurely exploration. But the arboreal primate's first encounter is likely to be more traumatic. A young monkey dropping out of a tree in the Congo would have the fright of his life before surfacing again.

But even this high-diving experience is not a universally effective deterrent. One species of arboreal monkey, the talapoin, has learned to utilize the drop into the Congo as a defense strategy. These little monkeys sleep clustered together on branches overhanging the water; and Annie Gautier, of the University of Rennes, has observed that if any danger approaches from behind they promptly splash down into the river and swim away.

One species of monkey is so aquatic that Richard Mark Martin's *Mammals of the Sea* includes it in a special appendix of "Other species which visit the sea." This is the proboscis monkey, an expert swimmer and diver. Its natural habitat is among the mangrove swamps of Borneo though it has sometimes been observed, or even picked up, by boats a couple of miles out at sea. On more than one occasion an entire social group of proboscis monkeys has been seen swimming across a river in Sarawak.

Among apes, as opposed to monkeys, the aversion to water is very rarely overcome. If *Homo sapiens* ran true to anthropoid type in this matter he would not build himself swimming pools or immerse himself in the sea purely for pleasure as he does today.

2 Human Swimming

It is sometimes claimed that the capacity to swim and dive cannot be part of our evolutionary inheritance because these acquirements are culturally transmitted. According to this argument no one can learn to swim unless he is taught to do so.

Even if it were true, the argument is scarcely conclusive. No one would learn to speak unless he were taught to do so. Yet very few people nowadays would doubt that the capacity for speech is part of our evolutionary inheritance. When young fresh-water otters first encounter water, they have to be chivvied into it by their mothers and often display some resistance—yet no one deduces from this that the otter is not an aquatic mammal.

In fact, it is not true that human beings would not acquire the art of swimming unless they were taught. Anthony Storr describes the experience of one group of doctors in London: "The pioneer doctors who started the Peckham Health Center discovered that quite tiny children could be safely left in the sloping shallow end of a swimming bath. Provided no adult interfered with them they would teach themselves to swim, exploring the water gradually and never venturing beyond the point at which they began to feel unsafe."

3 The Diving Reflex

In tables comparing the diving feats of men and other animals, man is sometimes described as a "nondiving" mammal. Yet he dives deeper than a beaver, most species of otter, and some species of porpoise and dolphin. The larger cetaceans, of course, are in a different league altogether. But a walrus in the open sea very seldom dives deeper than 300 feet; and recently an Italian diver reached a depth of 262.5 feet in one held breath and without artificial aids. If a nondiving mammal is a mammal that cannot dive, it is hard to follow the line of reasoning that places man under that heading. (See figs. 8 and 9.)

It is true that efficient diving is a result of training: so is an efficient performance on the trapeze. But no man could ever have become a trapeze artist if his remote ancestors had never swung by their arms from the branches, and similarly no man could become an exhibition diver if he were shaped like a gorilla or a chimpanzee and had no more control over his breathing than they have.

Man is known to share some of the physiological diving adaptations of the aquatic mammals, though not to the same degree. He shares, for example, the "diving reflex." This mechanism is found in all aquatic mammals. (It has also been observed in nonmammalian divers such as the frog.) When an aquatic mammal dives, the diving stimulates the vagus nerve which, among its other functions, acts as a cardiac inhibitor. This has the effect of reducing cardiac output and slowing down the rate of the heart beat, with a consequent reduction in the consumption of oxygen. In the case of human beings, the heart rate during a deep dive is

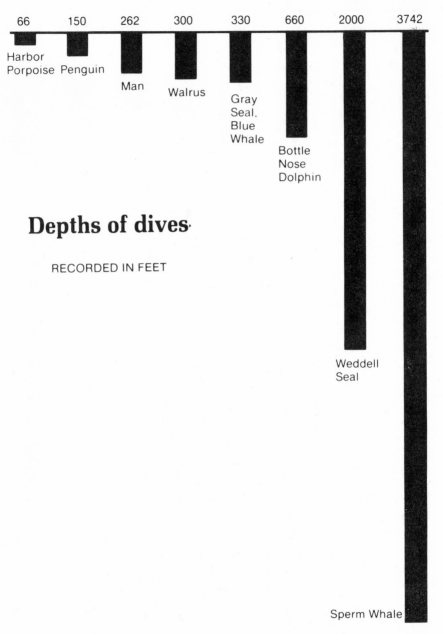

66   150   262   300   330   660   2000   3742

Harbor
Porpoise Penguin

Man

Walrus

Gray
Seal,
Blue
Whale

Bottle
Nose
Dolphin

Depths of dives.

RECORDED IN FEET

Weddell
Seal

Sperm Whale

Fig. 8.   If a nondiving mammal is a mammal that cannot dive . . .

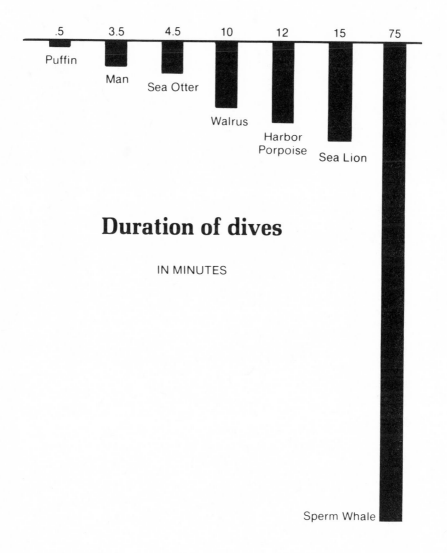

.5   3.5   4.5   10   12   15   75

Puffin

Man

Sea Otter

Walrus

Harbor
Porpoise

Sea Lion

Duration of dives

IN MINUTES

Sperm Whale

Fig. 9. . . . it is hard to follow the line of reasoning that places man under
that heading.

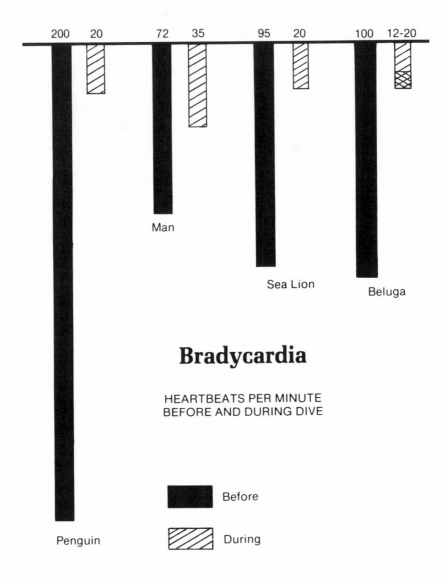

Bradycardia

HEARTBEATS PER MINUTE
BEFORE AND DURING DIVE

Fig. 10.   The diving reflex slows down the heart rate and reduces consumption of oxygen.

roughly halved. A comparison between man and some aquatics in respect of bradycardia (the slowing down of the heart rate) is illustrated in the graph (fig. 10).

Sudden immersion also produces some heart-rate reduction in many land mammals: it is the degree of bradycardia that distinguishes the aquatics. Yet to achieve a comparison between man and apes in this respect would seem to be impossible. No nonhuman primate, such as a gorilla or chimpanzee, can ever be induced to put its head under water, let alone to dive, except by force. Under these circumstances, the consequent exertion, aggression, and panic would affect the heart rate so as to render the experiment invalid.

4 Drowning

Jan Wind argues that while bradycardia in nonhuman mammals contributes to their survival in water, it can add to the drowning risk in man because he depends more for his survival on the cerebral cortex, which is said to be particularly sensitive to shortage of oxygen. Wind concedes the infant's greater ability to survive oxygen privation, but of human adults he makes the claim that "of various mammals, and maybe of all vertebrates, man is able to survive total immersion the shortest time."

This is a subject on which our ideas have been radically modified in recent years. *The Chicago Tribune* in 1977 featured a University of Michigan study on the effects of submersion. It reported several cases of prolonged immersion followed by no permanent ill effects. The most striking was that of an 18-year-old youth who was trapped underwater in a car for 38 minutes in Lake Michigan. On being brought to the surface he was resuscitated—a procedure that at one time would never have been contemplated even after a much shorter period of "drowning." The youth sustained no brain damage, and went on to become an A student in college. The incident led the National Oceanic and Atmospheric Administration to urge that no apparent drowning victim be given up for dead.

The youth had been trapped in very cold water, which almost certainly affected the outcome. The fact remains that dependence on an oxygen-starved cerebral cortex did not prove fatal over the space of 38 minutes.

Much more data would need to be supplied about survival times in other species before Jan Wind's description of *Homo sapiens* as the most drowning-prone "of all vertebrates" can be accepted as a scientific fact.

There is one further fact that makes it unlikely that man's large and complex brain renders him particularly vulnerable to drowning. Next to modern man, the largest brain (relative to body size) in any mammal species occurs in dolphins, and during their prolonged dives their brains continue to function perfectly well.

5 "Aquatic Animals Have Short Legs . . ."?

Attention is often drawn to the fact that in most aquatic mammals the hind legs have become relatively atrophied, and that in the most extreme instances of aquatic adaptation (the cetaceans) they have disappeared altogether. In man, on the other hand, the hind legs are relatively longer than those of apes.

Most of this increase in the relative length of human legs has taken place in the last four million years, i.e. between *Australopithecus* and ourselves. During the whole of this period bipedality has been fully operative, and the exigencies of this unique form of locomotion have favored the lengthening of the human stride and therefore of the legs.

In any case, the aquatic ape's physiological adaptation would differ in some ways from that of other aquatics because no other aquatic mammal is descended from a primate. This would affect, for one thing, his style of swimming.

Jan Wind points out how this primate heritage influences human arm movements in the crawl stroke: "Man owes the relatively high speed attainable by this type of swimming to his arboreal ancestry, facilitated by his highly moveable shoulder joints, associated with primate suspensory function" (i.e. monkeys and apes hanging from the branches). "Man can, in contrast to most other vertebrates, lift his forelimbs above the water level, thereby reducing his resistance."

The same factor would affect his leg movements. The legs of most aquatic mammals became attenuated because they are descended from quadrupeds of the standard construction whose hind limbs, like those of a dog, move only in one plane—backward and forward. This movement has very little propulsive power in the water.

But an ape can swing his legs, like his arms, outward at right angles to his body; he can also flex them at the knees. This position is reminiscent of (a) a swimmer doing the breast stroke; (b) the position most readily adopted by a young baby placed face downward on land or in water; and (c) the swimming action of the frog. This type of leg movement has appreciable propulsive power in water, and there would be no reason why the swimming ape, any more than the swimming frog, should find its legs dwindling as a result of its aquatic habitat (see fig. 11).

Dr. Robert D. Martin has pointed out, in his 1979 article on prosimian evolution, that a crucial feature throughout primate evolution has been increasing emphasis on hindlimb domination, as opposed to the forelimb domination characteristic of terrestrial (nonarboreal) mammals. This could be a perfectly adequate explanation of why an aquatic primate would automatically emphasize leg action, rather than arm action, when seeking an appropriate method of locomotion in water.

Fig. 11. Not all swimmers have short legs.

6 "Aquatics Close Their Nostrils . . ."?

One of the commonest signs of aquatic adaptation is the ability to close the nostrils in order to keep out water. (It is not exclusively aquatic: camels can close theirs in order to keep out sand.)

Human beings cannot do this; but if their noses are compared with those of any other anthropoid, it might appear that they have taken some tentative steps in that direction. They have evolved "wings" around the openings of the nostrils, and these are supplied with muscles such as no ape possesses. Nowadays the nostril muscles are used only for flaring the nostrils during moments of anger or passion, and it is sometimes argued that this is contrary to the way in which the seal, for example, functions.

But in fact the muscles in the seal's nostrils operate in exactly the same way as our own. He uses the muscles only to open the nostrils when out of water and not to close them. The only difference is that when the muscles are relaxed the seal's nostrils are fully closed, whereas ours are not.

The human nose is one of humanity's most baffling hallmarks. It has been suggested:

(1) That it evolved to add resonance to human speech. Yet the most resonant of primate cries is produced by the howler monkey, which has no such appendage.

(2) That it evolved to warm the air before it reaches the lungs. No one has suggested any reason why man or any of his ancestors should have possessed more sensitive lungs, or have been breathing chillier air, than any other animal on the savannah or off it.

(3) That the human nose is not large, it is just that the rest of the human face is small. Neotenists are particularly bothered by this aspect of our facial features, because the prominent nose and chin in human beings quite destroy the otherwise almost perfect "juvenile-ape" aspect of the human countenance. They get around the problem by suggesting that our distinctive profile has not been produced by the nose and chin growing bigger, but by the bones of the rest of the face becoming retarded in growth, thus leaving these features fortuitously projecting.

The human nose cannot be written off as a leftover piece of skull that mysteriously neglected to become retarded. It is buttressed right down to its tip by a specifically evolved and clearly purposeful strip of cartilage. A possible function is suggested by the fact that when a human being dives head first into water, the flesh-and-cartilage roof over his nose effectively deflects the water from being forced up into his sinuses at high pressure. Children, *jumping* into a pool from a high level spring-board, hold their noses with their fingers. Diving head first they have no need of this precaution.

The only other primate structure that comes anywhere near to resembling the human nose is the schnozzle of that semi-aquatic primate the proboscis monkey. This feature, while modest enough in the female, is highly impressive in the male.

7 "Aquatics Have Webbed Feet . . ."?

Alister Hardy pointed out (Appendix 2) that webbing between the toes still occurs in a minority (around 7 percent) of human beings in all cultures.

Webbing between the fingers is rarer, except for the vestigial triangle of thin skin between the finger and thumb of the human hand. It is this triangle that prevents us from making an angle of much more than 90 degrees between the two digits, a feature not found in any other primate.

Where the webbing is more pronounced (Plate 4) it tends to be regarded as a disfigurement, and in our culture is usually removed as a measure of cosmetic surgery.

Congenital abnormality can take many different forms. Most commonly it is the absence, or distortion, or incomplete development, or reduplication, of some item in the standard pattern of the human physical structure. It is extremely rare for congenital abnormality to take the form of adding a feature (as here, the interdigital webbing) that is usually believed to have been absent from our own species and from *our whole biological order* (the Primates) throughout its evolutionary history. It is even stranger when the added item is one that is normal and functional in aquatic mammals, aquatic birds, and aquatic reptiles—but in no non-aquatic species.

VIII
Aquatic Babies

1 The Swimming Infants

In the 1960s, when Hardy first proposed his theory, some people could not accept the possibility of aquatic apes because they regarded the aquatic environment as far too dangerous for any primate infants. It would be years, they felt, before the babies were old enough to learn to swim.

In recent years that piece of conventional wisdom has been turned completely upside down. It has been discovered that human babies are able to swim not merely before they are able to walk, but before they are able to crawl. The mistake made in the past has been not in introducing them to the water too soon, but in delaying it too long.

One Los Angeles swimming instructor said that at a very early age swimming comes naturally to babies, whereas if they have to learn it after they are over ten months old "... it is as if they had forgotten how to do it."

It should also be borne in mind that most of these training sessions are held in fresh water swimming pools. Sea water gives greater buoyancy.

The significant point is that babies in the first year of their lives appear to be quite as happy with their heads under water as above it.

They behave placidly, gazing around in the water with wide-open eyes and no signs of struggling or fear. (See Plate 6.)

Not long ago Charles Ramsay, swimming organizer for the Edinburgh County Council at the Royal Commonwealth Pool, was interviewed about his program for mother-and-child swimming lessons. He said: "During the first year of their lives they have an inborn reflex which stops them breathing during short spells under the surface. As a result no water can get into their lungs. They have remarkable breath control which they lose when they are over a year old. Nobody can explain why. But at this age they will not cough or panic under water.

"They also have a natural buoyancy because of their fatty tissue, which they lose when they start to crawl or walk. But at the age we get them, they have no fear of water. It is virtually impossible for them to come to any harm even when they are placed face down on the surface of the pool."

Similar programs are being introduced in many different countries. Dr. Igor Tjarkovsky of Moscow reports that his three-month-old daughter could hold her breath under water for three minutes without ill effects.

In no case have there been any reports of ill effects on the children. On the contrary, it has frequently been claimed that they benefit from it. Children initiated thus early into water are said to display a greater than normal degree of independence and confidence in themselves as they grow older. However, one cannot place too much reliance on these claims without further evidence, since up to a point the parents themselves are self-selected for confidence and independence—otherwise they would not volunteer themselves and their children to take part in the experience.

2 Fatty Tissue

A reference to the fatty tissue that human infants lose when they begin to walk or crawl may have some bearing on one of the many anomalies that comes to light on comparing human and ape reproduction.

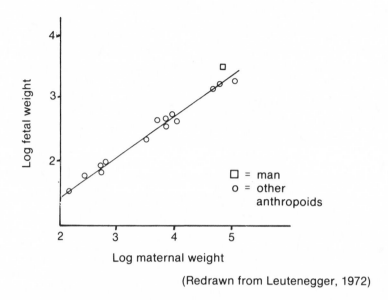

(Redrawn from Leutenegger, 1972)

Fig. 12.   Compared with all the other anthropoids, the human baby is born "too heavy," relative to the mother.

In 1972 Leutenegger constructed a graph plotting the weight of the fetus against maternal weight for 15 monkey and ape species. The weight of the babies varies with the weight of the mothers in a fairly predictable fashion for 14 out of the 15 species. The one obvious exception is the human baby. If the human mother in the series had conformed to this pattern, then the predicted birth weight of the baby would have been around 2200 grams (4½ lbs.), as in the great apes. The actual weight was around 3300 grams (7 lbs.). Compared with all the other monkeys and apes the human baby is born "too heavy" relative to the weight of the mother.

The greater weight of the human infant cannot be explained by the fact that a human baby has a longer gestation period. A graph drawn by Hugget and Widdas in 1951 plotted birth weight against length of gestation period; and once again the human baby failed to conform to the primate norm. The graph showed that it is significantly heavier than would be predicted from the length of its gestation period.

This human peculiarity has been attributed to neoteny. In the offspring of monkeys and apes the fetus grows fast in the early stages, but the rate of growth slows down as birth approaches. Neotenists argue that in human babies the rapid fetal growth rate is retained up to and past the time of birth, owing to retardation.

If this is indeed the reason, then we would expect that this continued growth would apply to all parameters equally: that the baby would continue to grow at the rapid fetal rate in height, for example, as well as in weight. Measurements have so far failed to establish this point.

What actually happens to the human baby as birth approaches is somewhat different. Rather than a continued rapid rate of general growth, there is the rapid laying down of the subcutaneous layer of fatty tissue that is so singularly lacking in the newborn chimpanzee. This happens to no other primate fetus. It would make sense if it were a preparation for launching the newborn infant buoyantly into a watery environment; it would add rapidly to weight (in air), while adding nothing to other parameters such as size or maturation of the skeleton— and it would explain the "anomaly" in the graphs.

3 Underwater Childbirth

Launching the newborn straight into the water may seem an alarming procedure even now that we have learned so much about the ability of young babies to be happy and safe in the element.

But numbers of babies have been born that way in the clinic of Russian gynecologist Dr. Igor Tjarkovsky. (See Plates 7 and 8.) After birth, the newborn is guided gently to the surface by the midwife to take its first breath. Incidentally, this is what frequently happens at the birth of a dolphin. A dolphin "midwife" attends the birth, ready to nudge the baby to the surface in case anything should hinder the mother from doing so. Richard Mark Martin records:

"The reproductive life of dolphins is especially well studied, thanks largely to the magnificent American marine aquariums. . . . Gestation takes about a year. A pregnant dolphin will often keep a little distance between itself and the remainder of the school as the moment of birth draws near; it will sometimes choose a companion female, who performs the functions of a midwife and who remains in close attendance."

Dr. Igor Tjarkovsky first embarked on his experiments following the birth of his own daughter, who was born prematurely, weighing little more than two pounds, and was regarded by doctors as a hopeless case.

Because, as he explained, he believed in the exceptional healing effect of the water environment, he put his little daughter into a warm bath—"thus, in fact, returning her to a world where she was safe. . . . It would not be an understatement to say that she spent the first two years of her life in water. Very soon we had no more fears regarding her condition, and it was obvious that she was far more advanced, both physically and intellectually, than most children of her age. . . ." He also stated that when she was three months old she could manage to stay under water without taking a breath *for up to three minutes,* with no ill effects.

When he decided to investigate the possibilities of the actual birth taking place underwater, he enlisted in the first place the cooperation of top Russian swimmers, on the grounds that since water was a natural element to them, they would be the ideal subjects.

At one time he claimed to have successfully delivered babies not only in obstetric tanks but beneath the surface of the Black Sea.

At first, the authorities were cautious about his experiments, but later the conviction appeared to grow that there was some value in his theories, and he was encouraged to teach his techniques to others. He claimed that babies delivered by this method grew up to be healthier and more intelligent, and that the mothers found the method to be relatively free of pain and stress.

This, like the report of psychologically advantaged "water babies," must be regarded as a subjective impression. A mother's euphoria might be enhanced by the exhilaration of taking part in an unusual experiment, or by receiving more concentrated attention than she might receive in the average labor room. Alternatively, the need to concentrate on coordinating her movements in a three-dimensional medium might divert her mind from any less pleasant sensations.

An official statement from the Consultative Council on Obstetrics and Gynecology, USSR Ministry of Public Health, has subsequently reassessed Dr. Tjarkovsky's experiments and pronounced them to be "baseless and dangerous," and they have been discontinued. But the photographs appear to confirm that for some of the mothers and babies involved the experience was entirely free from trauma.

And while in the Soviet Union official backing for underwater childbirth has been withdrawn, in the West interest is growing. One proponent is Dr. Michel Odent, a French surgeon who gives women a free choice of the position in which they wish to give birth—for example, standing or squatting instead of supine. He has also equipped his hospital with a pool of warm water for underwater deliveries, and mothers frequently opt for this method. Odent's pool, unlike Tjarkovsky's, is shallow enough for the mother to sit in. Patients report that the method is stress free and relaxing.

It is not being suggested that the aquatic ape was ever so fully aquatic that underwater childbirth was the norm. Alister Hardy's initial concept certainly envisaged a littoral habitat and a much more moderate degree of aquatic adaptation. More research will be needed, together with more detailed comparisons of the behavior and anatomy of both primates and aquatic mammals, before questions of this kind can be satisfactorily answered.

One fact about mammal childbirth may be worthy of note in this connection. Terrestrial mammals, including herbivorous species and nonhuman primates, consume the placenta after birth. Fully aquatic mammals are unable to do this because it sinks to the sea bed so that they have lost the instinct to do so. It would seem that *Homo sapiens,* even in the most primitive and cannibalistic of societies, has also lost this instinct.

4 Postnatal Responses

Once the baby has been born, it exhibits a variety of unlearned responses. Many of these we do not yet fully understand. There is, for example, a phenomenon known as the swimming reflex of the newborn, described in the *Nelson Text Book of Pediatrics.* "By nine weeks contralateral flexion may be followed by ipsilateral flexion (swimming motions)"—as though an initial instinct to bring the arms together and hold on to something were succeeded by a later instinct to paddle around, as today's water babies seem to do.

There may also have been an adaptive advantage in the "breath holding" of babies—a phenomenon that occurs only up to the age of 2 years. Young children quite often react in this way to fear or resentment or an imagined danger of being deserted. Harrison (1960) contends that human infants are specially adapted for breath holding by a proportional enlargement of the vertebral foramina.

The breath holding may be prolonged until the parents are seriously frightened. If the experience is repeated, a child may learn to use breath holding as a kind of emotional blackmail, and then the whole syndrome is written off as "being naughty." The first manifestation of it, however, is always purely instinctive and uncontrolled, and no explanation of the phenomenon has yet been advanced.

Dr. T. B. Anderson of Cambridge has speculated that asthma (unknown in apes) may be connected with a one-time biological adaptation. He points out that in a diving seal the bronchial sphincters are constricted, and draws attention to a similarity in the paravertebral venous drainage of seal and *Homo sapiens.*

Martin Nemiroff, a specialist in diving medicine at the University of Michigan, stresses that "the younger we are, the more active is the diving reflex. It may be one of the protective mechanisms that allow us to survive birth."

5 Non-Aquatic Explanations

Neither savannah nor neoteny theories suggest any way of accounting for these phenomena.

The only non-aquatic explanations that have been advanced attempt to link them with prenatal experience of a watery environment in the uterus. One anthropologist used this argument when confronted for the first time with Hardy's contention that the direction of the hair tracts in the coat of hair (lanugo) covering the unborn fetus resembles the passage of water over a swimming body. (Appendix 2) He hazarded that the fetal hairs took up this position because the unborn child had been "swimming around and around in the amniotic fluid." In fact, human fetuses rotate only very slowly; they do not differ from other mammals in this respect, and this factor could not possibly explain why the hair tracts should differ.

A similar line of reasoning attempts to account for the infant's contented acceptance of submersion in water. It is suggested that the babies are happy because they imagine they have returned to the secure world they inhabited while still in the uterus. Secure it may have been, but there cannot be much resemblance between the subjective sensations of being packed into a very confined space in a fetal crouch and floating free in a swimming pool. (See fig. 13)

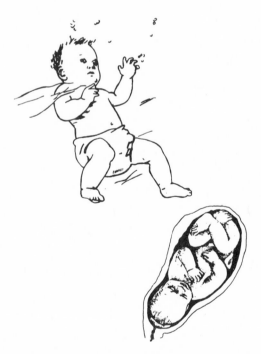

Fig. 13.   The swimming pool can hardly be said to recreate the conditions of prenatal existence.

This kind of argument breaks down on two counts. One is that the fetuses of all unborn mammals—kittens and puppies and foals and leverets as well as human babies—are suspended in precisely the same kind of amniotic fluid before birth. Yet there is no evidence that they carry the memory of it into the world outside, or that it gives them any kind of affinity for an aquatic environment.

Secondly, it is very hard to see how prenatal experience can have any bearing on the breath holding or the diving reflex or the tolerance of oxygen privation. Before birth the child's lungs have not been brought into operation at all: it cannot have learned the knack of inhibiting a function it has never performed. And there would be no point in acquiring the ability to tolerate anoxia. In the uterus the fetus's oxygen supply still comes to it through the umbilical cord, while once outside the uterus, oxygen is uninterruptedly available from the air—for the young of all non-aquatic mammals.

1 Apart from man, the only sea-going primate is the proboscis monkey. This one, observed swimming a couple of miles out in the China Sea, was glad to accept a lift home.

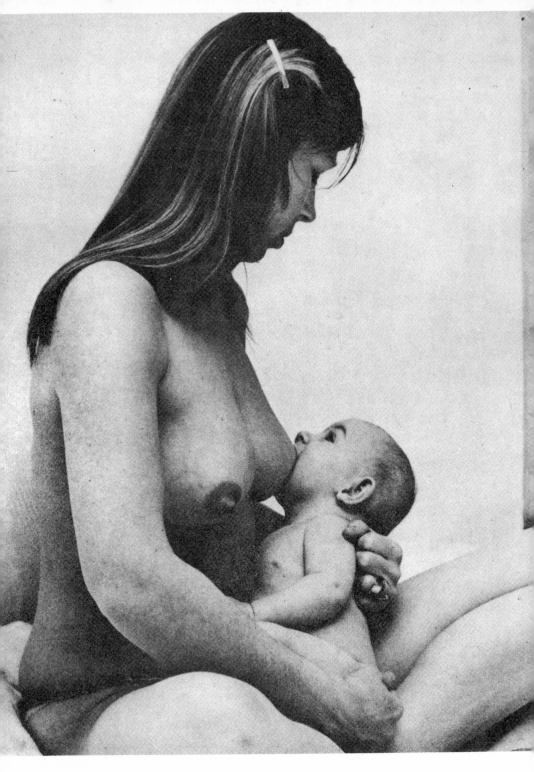

2 The human baby is plump and naked; the chimpanzee's is skinny and covered with

hair. Scientists unable to explain these differences dismiss them as "merely relative".   3

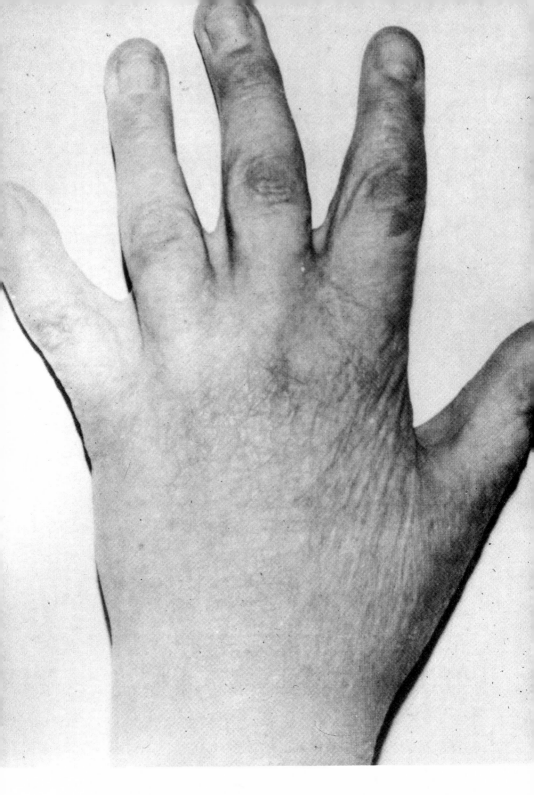

4 Interdigital webbing most often occurs between the toes, but sometimes the hands are affected.

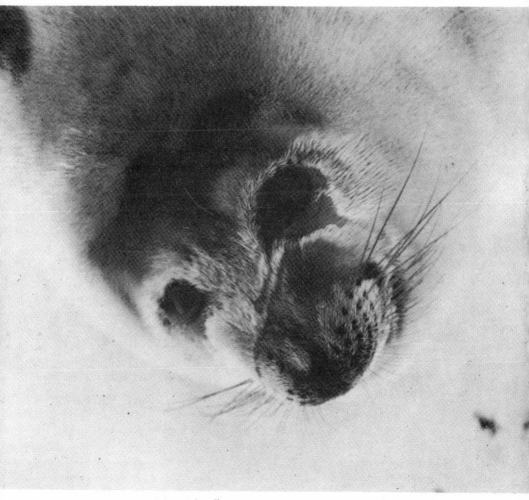

5 "Then the tears flow thick and fast."

6 Human babies are able to swim before they are able to walk.

7 Numbers of babies have been born that way.

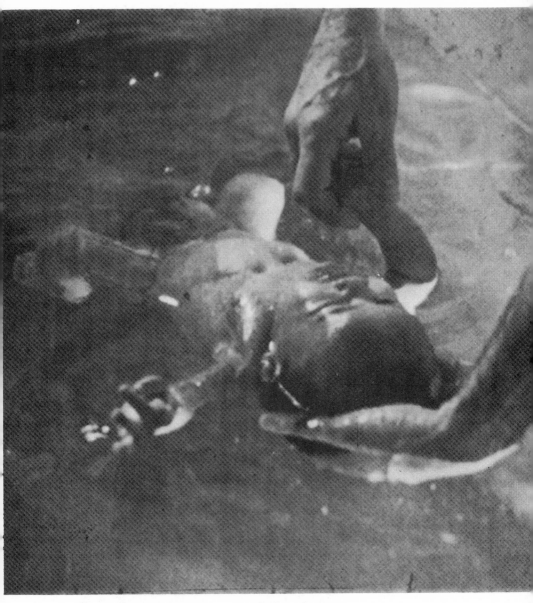

8 It is clear that the experience can be free of trauma.

IX
Speech

1 An Intractable Problem

The capacity to speak is one of the three major hallmarks of humanity. Even more dramatically than bipedalism and the use of tools, it sets us irrevocably apart from all the rest of the animal kingdom. The question of how and why an anthropoid began to speak is central to all our efforts to understand man and his evolution.

It may be the most important question, and it is certainly the most intractable. Anthropology, like every other science, deals in facts and measurements. Yet as soon as it tries to deal with the origins of speech, many of its most valuable tools become useless.

Field work in archeological sites can have no direct relevance here. By digging up fossil remains, measuring skulls and limbs and carbon-dating bones and teeth, a few hard facts about the evolution of bipedalism may be established. By locating and studying prehistoric artifacts, valid statements can be made about the beginnings of durable toolmaking. But speech is a behavior pattern—and behavior does not fossilize.

Comparisons with other species are often very illuminating, but no one can draw graphs illustrating the evolution of speech. There is no series of primates at varying stages of speech acquisition with which we can usefully compare ourselves.

The possibilities of experimentation are circumscribed. Experiments can be envisaged to discover whether a group of freely interacting nonhandicapped children would evolve a vocabulary for themselves if they were not supplied with one—but the species involved is our own, and its infants cannot be used as guinea pigs.

Four approaches to the problem have been employed: speculation, a study of the modes of communication used by the primates, comparative anatomy, and attempts to teach chimpanzees to speak.

2 The Savannah Theory

In popular expositions of the savannah theory, the problem of speech is dealt with speculatively. It takes the form of pointing out:

(a) that the savannah hominids would need an improved system of communication when they embarked on toolmaking and the cooperative hunting of big game;

(b) that vocal communication turned out to be an excellent way of fulfilling that need once it had passed over the great divide between involuntary animal noises and meaningful speech.

However, observation (b) is based purely on hindsight. No animal embarks on a mode of behavior because it is going to be of immense benefit to its descendants a million years later.

If such speculations are to have any validity they must be based on the options open to the hominid *before* speech was developed. These can be deduced from the modes of communication that primates had been utilizing and improving upon throughout the whole course of their evolutionary development.

3 Primate Modes of Communication

Communication between primates is conducted by means of scent signals, touch signals, vocal signals, and visual signals.

Scent signals serve to communicate identity and states of mind and body such as anger, fear, or sexual receptiveness. They promote bonding between individuals and groups by enabling a mother to recognize her own infant, and enabling an animal to distinguish instantly a member of its group from an outsider. They are also used to establish territorial claims.

Touch signals in primates serve to cement the mother/child bond by holding and cuddling and suckling. They establish socialization among the growing young by rough-and-tumble play behavior. They establish hierarchies of dominance and submission by ritual presenting and mounting behavior between males as well as between males and females. Peaceful interactions of all kinds are expressed and reinforced by frequent grooming sessions indicative of harmony and amiability.

Vocal signals in primates, like olfactory ones, are largely involuntary expressions of states of mind—panic, rage, grief, protest, frustration, appeal, alarm. Among primates—as among birds and other mammals—an alarm call besides expressing a state of mind also conveys the information that there is danger. Variations in the kind and degree of alarm expressed can sometimes convey the type of danger that has been perceived. For example, Leonard Williams observed that one particular alarm call is used by woolly monkeys only when a baby monkey has fallen from a tree.

But between primates, especially anthropoids, visual communication is the channel that has been developed to the highest degree of precision. Through the medium of a great variety of gestures, postures, movements, facial expressions, and the management of spatial relations between individuals, they can convey to one another with considerable sublety their feelings and wishes, their immediate intentions, and their social relationship.

4 The Choice of a Channel

Given that the savannah anthropoids were thus endowed with four channels of communication, it is hard to see why either the exigencies of hunting or toolmaking would make it likely that they would choose the vocal one—still in a primitive state of development—rather than the visual one that was so highly developed.

Many different carnivorous species—from wolves and hyenas to prides of lions—have learned to cooperate in hunting. None of these species has found it necessary to converse about it.

Among primates, baboons have evolved elaborate military-type strategies for defense on the savannah against prowling predators: e.g. the females and the young in the center and the most dominant male outriders on the periphery. In *Social Groups of Monkeys, Apes and Men,* Michael Chance and Clifford Jolly describe how different species of baboons deploy their forces in the face of danger. This is all achieved without the aid of the spoken word.

Among African hunter/gatherers the !Kung bushmen are able to talk, but they do not make use of this talent on the hunt. Silence and stealth are vital, and the hunters confine themselves to visual signals until after the kill.

As for the cultural transmission of toolmaking, it may not be easy to give a visual demonstration of how to knap a flint, but it would be a hundred times more difficult, even given modern day standards of verbal dexterity, to convey the knowledge verbally.

5 Comparative Anatomy

Such speculations do not really bring us nearer to an understanding of why men speak and apes do not. The earliest attempts to establish some hard facts that would throw light on this were in the field of comparative anatomy. A search was made for anatomical differences between man and the other anthropoids that might explain their failure to follow our eloquent example.

Very little light was thrown on the matter. It was found that human vocal cords are somewhat blunter on the edges than those of apes and baboons. Bipedalism has altered the way a man balances his head, so that his larynx has slid a little farther down his throat. This has created a long tubular resonating cavity that enables him to produce low-pitched vocal sounds. His jaw is shorter, bringing his front teeth nearer to the back of his throat. There are some modifications in the muscles around his eyes and mouth.

Another particular feature of human anatomy is the ability to open or close the nasal passages during the production of speech by the operation of muscles that can raise or relax the velum (see fig. 14).

Vocal tracts of adult human

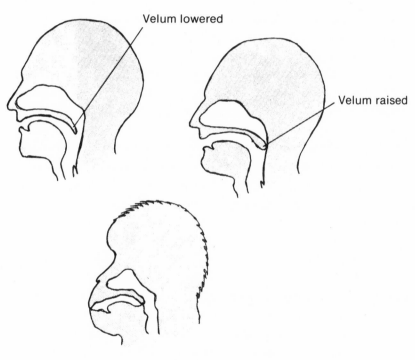

Velum lowered

Velum raised

Vocal tracts of orangutan

(After Negus, 1949)

Fig. 14. Human beings have a mechanism that allows the nasal cavity to be disconnected from the other air passages.

Philip Lieberman, in his book *The Origins of Language,* states: "Humans have a speech-producing mechanism that allows the nasal cavity to be either connected to or disconnected from the other supralaryngeal air passages."

It seems highly probable that this capacity may have evolved for the convenience of the aquatic ape—to prevent any water that entered the nasal cavity from reaching the lungs. The velum adopts the closed position during diving or swimming underwater: this is the procedure known as "holding one's breath." The capacity to make this adjustment has certainly been exploited to increase the variety of sounds by producing "nasal" consonants in speech, but that may well have been a secondary use. The teeth and tongue are used in speech production, but no one argues that they evolved for that specific purpose.

All these differences between our anatomy and that of the ape are comparatively trivial—certainly not enough to keep the ape tongue-tied. He can, and does, produce noises equivalent to many of our vowel sounds. He can smack his lips or round them into an "o." He can click his teeth and make the sounds of "k" and "p." He can and often does produce a whole series of glottal stops. He can gibber and whimper and squeal and chatter and moan and make a noise not unlike laughter. The differences in the larynx and the vocal cords might have been sufficient to render the speech of the chimpanzee hoarser or less distinct than our own, but there was nothing in any of the organs on the dissecting table to explain why we could say so much and he could say nothing at all.

Encouraged by the lack of any obvious anatomical incapacity in the primate's vocal organs, scientists began to try to teach chimpanzees to speak.

6 Viki and Washoe

Two American psychologists, K. J. and Caroline Hayes, taught their chimpanzee Viki to pronounce four words not very distinctly: "papa," "mama," "cup," and "up." It took all three of them six years of unremitting effort to produce this result.

Later, Allen and Beatrice Garner decided to try to teach their chimp Washoe sign language. They reported that within two years she had acquired a vocabulary of 34 signs, and was rearranging the signs to form new combinations, such as "open-food-drink"—which may have been a compound for "refrigerator," or a full-scale sentence meaning, "Open that door and give me some milk and a banana."

Since then attempts have been made to teach sign language to chimpanzees and, more recently, to gorillas and orangutans. They have met with varying degrees of success, but the apes undoubtedly find communication by signs easier and more congenial than Viki found her training in the spoken word.

After all, visual communication is the chimpanzee's long suit. In the wild, if a chimpanzee sees another with a coveted tidbit—especially meat—he will hold out his hand with palm extended, which means, "Give me," in the language of any higher primate, including *Homo sapiens*. Apes can and do readily imitate our movements and gestures: that is why "to ape" means "to imitate." They can understand many of our verbal concepts, as Washoe proved, but they find it overwhelmingly difficult to imitate our spoken words.

There must have been a time when our own primate ancestors would have had the same overwhelming difficulty. Something very unusual must have happened to make us switch over to vocal communication. Either it became easier for us, or more essential to us, or both.

7 The Aquatic Experience

Among aquatic animals communications by sound is of paramount importance. Among the most highly adapted—the cetaceans (whales and dolphins)—auditory perception is so dominant that it has even usurped some of the functions of sight, with objects in the environment being perceived by echo location.

The reason for the change is not too far to seek. When a land-dwelling mammal moves to an aquatic environment, the operation of several of his normal modes of communication is disrupted.

Scent signals lose their usefulness. He has been designed to detect them by inhalation of scent-laden air, which cannot be inhaled under water. Visual signals become far less practicable. On land, with his feet on the ground and his companion's eyes fixed steadily upon him to divine his intentions, every move of hand or eyebrow, every minute postural shift, can be loaded with meaning. In water a high proportion of his postures and movements are dictated to him by the need to keep afloat or swim or dive. Minute changes of facial expression are harder to perceive and decipher under water than they would be in the clear air and bright light of the savannah. For a swimming aquatic it is very much harder to maintain eye-to-eye contact with his companions and receive subtle incoming messages in that way. Leisurely grooming sessions are out of the question.

Thus, if there is any shift in the relative dominance of the various senses between apes and ourselves, this might provide some evidence as to what happened to us during the fossil gap.

8 That Sense that Atrophied

To say that our sense of smell diminished may seem to have little significance. This, after all, is only a continuance of a trend that has been a constant feature of primate evolution.

Primates are tree dwellers. Scent trails are much harder to follow through the tree tops than when imprinted on the damp earth. R. D. Martin (1979) demonstrated, in comparing nocturnal with diurnal primates, that as various primates became specialized for daytime activity, their olfactory lobes—the section of the brain devoted to scent perception—shrank. Arguably, it would not be surprising if in man the lobes had continued this unwavering trend and shrank a little further.

(It could be argued that it *would* be surprising, if man's ancestor was merely a grounded savannah ape. It might be expected that his sense of smell on leaving the trees would be revitalized. Birds in general have little sense of smell, but the flightless ones, like farmyard ducks, tend to recover it to some extent.)

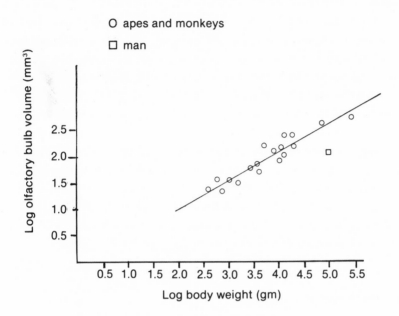

Fig. 15.   Man's sense of smell has diminished more sharply than would be predicted. (After R. D. Martin, 1979)

However, man did not merely continue the unwavering trend—he sharply accelerated it. The graph in figure 15 plots the size of the olfactory lobes in the brains of monkeys and apes against body size. The graph shows that, just as the babies of *Homo sapiens* are born heavier than would be predicted, so his sense of smell has diminished more sharply than would be predicted.

Contrary to popular legend, this has not happened in our own time as a consequence of civilized life, deodorants, and gasoline fumes. It is part of our evolutionary inheritance that has not yet been explained.

In all aquatic mammals the size of the olfactory lobes, with respect to body size, is diminished. In whales and dolphins the lobes have vanished altogether.

9 Hearing and Uttering

Our sense of hearing has not grown correspondingly more acute, but it has become more specialized. It is adapted to the function of listening more attentively to one another, even at the cost of losing some incoming messages from the environment. A chimpanzee can hear tones up to 33,000 Hertz, whereas a human being can normally hear only up to 24,000 Hertz; but our hearing is extremely sensitive within the range of the human voice.

However, the most remarkable human development in the field of vocal communication is that we have acquired *conscious control* over the utterance of sound.

This has been achieved together with the conscious control of breathing, which is a feature of, and essential to, all diving animals. It is highly unlikely that the two developments are unconnected.

Viki found it so hard to learn how to speak because animal utterances are involuntary and stereotyped calls. The clucking of the hen or the crowing of the rooster may carry some informational content to the listener: "An egg has been laid" or "The sun is rising," but these utterances are unconditioned reflexes, not voluntary attempts to communicate. If the rooster is castrated, he will never crow, however often the sun rises.

Recently there has been considerable discussion of the fact that the vervet monkey, a savannah dweller, has three different alarm calls: one for "snake," one for "eagle," and one for "leopard." When these alarm calls are taped and played back to a group of vervets they will respond appropriately—by searching the ground around them, diving under a bush, or climbing a tree. To savannah theorists this is the beginning of true speech.

But this cannot be so. The vervet possesses the inherited potential for the three cries as a bird possesses the inherited potential for "caw" or "cuckoo," or a cat the potential to mew or purr in response to the appropriate stimulus. There is no evidence that the vervet can produce the calls in the absence of the stimulus, or inhibit them in the presence of the stimulus.

Viki was being asked to acquire control over a normally involuntary process. It is as though a human being had been requested to demonstrate voluntary control over blushing, or the rate of his heart beat, or the dilation of his pupils, or the processes of his digestion. With patience and determination, he can gain at best a partial degree of control over some of these mechanisms—just as Viki, in her strong desire to please, acquired some very minor mastery over her lungs and vocal cords.

If a prehuman hominid, in the course of learning to swim and dive, had acquired conscious control over breathing and vocalization for the first time in primate history, then the overwhelming difficulty hampering the development of the vocal channel would have been removed. He would have returned to the land not necessarily articulate, but equipped with the indispensable preadaptation for speech. He would, like us, have been able to control his breathing as he controls his limbs, so effortlessly that he never gives it a thought. Without that capacity, controlled phonic utterance is impossible.

10 Controlled Vocalization by Aquatics

The most powerful conditioning agent yet discovered by biological researchers is a fine wire capable of applying electrical stimulation direct to the "pleasure" centers of an animal's brain. A rat that is enabled to afford itself this stimulus by pressing a lever will continue pressing it until it is exhausted. Animals can be trained to perform complicated maneuvers with the incentive of obtaining this reward. But terrestrial laboratory animals such as rats and rhesus monkeys have proved unable to learn to attain it by uttering any kind of sound.

On the other hand, aquatic animals such as circus sea lions can be trained very easily to utter sounds to order, with no stronger incentive than the reward of a piece of fish or the trainer's approval. The same is true of dolphins.

Some aquatic animals make use of their talent for voluntary sound production to obtain information about their surroundings by means of echo location. The dolphin's famous talent for echo location is often discussed as though it were unique. But this is by no means the case. Whales use the same mechanism. And a totally unrelated species, the Antarctic Ross Seal, is known to use echo location in searching out cephalopods and soft-shelled crustaceans on the sea floor.

The other echo location specialist is the bat. His incentive for switching from visual to vocal signals would not have been in essence different from the dolphin's. Olfactory communication channels would be less useful to him than to a ground-dwelling shrew; he would find, like the birds, that he could not effectively scent-mark the air he was flying through. Visual signaling would be less useful because in flying, precisely as in swimming, the limbs are needed for other purposes than gesturing, and posture is determined by the exigencies of locomotion.

11 Speech: The Only Other Claimant

Opinions differ on the question of whether true speech has yet been achieved by any creature other than ourselves. But there is agreement on one point: if any living animal can be said to have achieved it, the likeliest candidate is the dolphin.

Dolphins sometimes try to imitate the sounds we make—(it has recently been claimed that seals occasionally do the same)—but they produce no kind of speech that is intelligible to us. It would be extraordinary if they did. But there are certain unusual features about the way dolphins communicate between themselves.

Among many studies of dolphin sound production made in recent decades by American scientists, one possibly significant observation was made by the Caldwells, a husband and wife team. Dolphin vocalizations are divided into two main groups: (1) pulsing sounds, which are chiefly for the purpose of echo location, and (2) whistling sounds. The Caldwells observed not only that the whistling of one animal in a school stimulates the others to do the same, but that "response whistling" does not begin until the first animal's whistling is completed. If two animals begin whistling simultaneously, one of the animals will often stop and wait until the other has finished.

Too much significance should not perhaps be attached to this, but it does seem to suggest that they are not merely *hearing* one another but *listening* to one another.

There remains the question of what kind of information they are conveying. Richard Mark Martin, in his book *Mammals of the Sea,* describes the kind of experiment designed to assess the effectiveness of vocal communication within the species.

"The dolphin's incredible powers of communication—indispensable in the wild—have been demonstrated in a variety of often spectacular ways. Two dolphins—'A' and 'B'—completely visually isolated in separate pools, were able to communicate with each other and even relay information. Dolphin 'A' was able to tell dolphin 'B' the correct lever to manipulate in order to obtain a reward of fish after the information had been given by the experimenter only to dolphin 'A.' Many tasks and variations on this theme were accomplished by the dolphins with so much assurance that it was obvious they did indeed have a language of sorts."

These claims on behalf of the dolphin are still regarded as controversial. Dr. Jarvis Bastian, who has himself conducted some of these experiments, warns that caution must be exercised in making deductions from them. But our knowledge of these playful, intelligent, and friendly mammals is increasing rapidly, and it should not be too difficult to design further experiments. These could preclude the possibility of error and finally establish whether or not we are the only species on this planet to have acquired true speech.

X
Toward a Synthesis

1 Recapitulation

Throughout the previous chapters the three main approaches to explaining the problem of human evolution (savannah, neoteny, and aquatic) have been discussed as though they were separate from one another and, in some sense, in competition with one another.

This impression, although misleading, has been hard to avoid because their respective proponents are apt to convey the impression that their own particular version contains all the answers. The Aquatic Theory, in particular, is often dismissed not by refuting or even discussing the arguments in its support, but by the assertion that "We don't need it. We can explain everything without that."

The aim of this book has been to suggest that this is too large a claim. If we exclude the aquatic dimension, we can explain a good deal about human anatomy and physiology—but by no means everything. And some of the non-aquatic explanations turn out on close examination to be specious and involved, whereas the aquatic explanations of the same phenomena are simpler, more consistent, and much more frequently supported by parallel development in species other than our own.

It is now time to attempt to bring the three schools of thought into a more organic relationship with one another.

2 Savannah and Aquatic

The savannah theory and the aquatic theory are in no sense mutually exclusive.

The long period of human development on the savannah, from the first appearance of *Australopithecus* onward, is established and proven beyond all argument. The familiar and well-attested story of the development of tools and weapons, the emergence of the hunting/gathering economy and the consequent emergence of new kinds of social relationships, is not challenged.

The aquatic contention is that other factors must have been involved. Many other primates have moved from the trees to the open plains, and in no single one of them has that move produced any of the changes that caused the ancestors of *Homo sapiens* to diverge so dramatically from all his nearest relatives.

Savannah theory alone cannot explain with any degree of plausibility the loss of body hair, subcutaneous fat, or tears. The striking differences between an infant chimpanzee and a human baby cannot be accounted for by the simple proposition that "Daddy's gone a-hunting."

Bipedalism, ventro-ventral sex, and vocal communication are very difficult to account for by savannah theory alone. Even the most orthodox of evolutionists have never been happy about the explanations commonly propounded.

Following an aquatic interlude—which would have introduced *pre-adaptations* to all these changes—it is easy to understand how in the subsequent millions of years on the savannah our ancestors would have continued to develop and to steer the aquatic hominid along the unique path to the emergence of man.

Moreover, if the aquatic ape, as is likely, was initially reluctant to wander too far from the water courses, then some of the aquatic influences would have continued to operate far into what we are accustomed to regard as the savannah period.

Many reasonable explanations are advanced to account for the fact that the earliest prehuman relics have almost all been found near water courses. The commonest explanation is that these conditions—the undisturbed sediments of the lakeside and river bank, the occasional flash floods causing the hominids to be entombed in the mud—were essential for the preservation of the earliest human remains. The implication is that *Australopithecus* existed in equal numbers far away from the rivers, but that conditions elsewhere did not favor the preservation of his bones. That is possible; but it is, so far, unproven.

It is equally possible that the fossils are so frequently found in these riverside and lakeside situations because they were the sites in which the hominids habitually lived.

3 Neoteny and Aquatic

Neoteny is a mechanism of evolutionary change—not an explanation for such a change. Whenever an organism exhibits this mechanism, there is usually a good reason for it. A sea snail adopts this strategy if the supply of calcium in its environment is not sufficient to build the thick shell of the adult form. By a process of neoteny it evolves into a juvenilized thin-shelled version of the normal adult. The axolotl *Ambystoma* resorts to neoteny when changes make conditions on land unfavorable for the survival of the adult form.

Neoteny, in fact, is a mode of responding to changing conditions. In discussing the process in primitive creatures neotenists often specify the kind of environmental emergency that "triggered off" the neoteny.

In speaking of man as the neotenic ape investigators have carefully refrained from speculating about the kind of change that triggered off the process. Yet a change there must surely have been. And a move to the savannah does not seem to produce that effect. The most typical and successful of savannah primates is the baboon—and *his* head, far from becoming flat-faced, juvenilized and neotenous, has burgeoned into a long snout and a dogfaced profile.

It might be worth considering whether there are any possible links between neoteny and aquatic adaptation.

4 Slowing Down

The particular kind of neoteny that has affected our own species is described as retardation. Compared to other mammals, and even to other primates, man moves very slowly through the stages of his life. He has a longer gestation period, a longer childhood, and a longer period of growth toward maturity. He also has a longer period of fertile reproductive adulthood and an extra period of postreproductive existence. His biological clock has slowed down more markedly than would be predicted from his body size (see fig. 16).

As a general rule in animal species, large animals tend to live longer than small ones; and the length of their life closely correlates with a number of factors. Their hearts beat more slowly; they take fewer breaths per minute. The rule is so constant that it has been possible to calculate that a mammal's expectation of life equals 200 million of its breaths, or 800 million of its heartbeats. (By this rule man is a nonconformer; he lives three times as long as he is "entitled" to.)

As a general rule also, the fossil records show that mammals adopting an aquatic existence tend to grow larger. The largest living animals are the great whales. The blue whale is bigger than the biggest of all the dinosaurs, *Diplodocus* (which was itself perhaps a marsh dweller). The largest land animals are the great pachyderms—aquatics like the hippopotamus or ex-wallowers like the rhinoceros. The life clocks of the marine mammals tick over very slowly. Where *Homo sapiens* when resting takes 15 breaths per minute, the Californian sea lion takes only six; the bottlenosed dolphin three and a half; and the killer whale less than one.

The slowness of life rhythms in aquatic mammals is usually attributed to their size, and their size in turn is usually attributed to the factor of heat conservation in cold water. The argument here is that the ratio of surface area to volume is lower in big animals than in small ones so that there would be a proportionately smaller body surface through which heat could be lost. It is hard to see how this factor would have produced the massive tropical hippo, who has more trouble keeping cool than

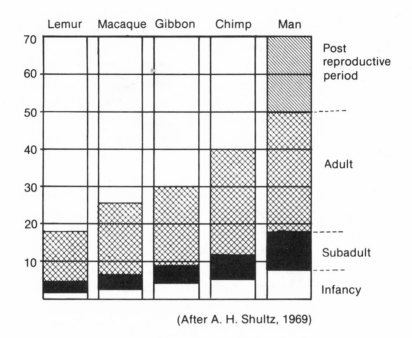

(After A. H. Shultz, 1969)

Fig. 16.   Compared to other primates, man moves very slowly through the
          stages of his life.

keeping warm, and the possibility of some other causal connection
between aquatic habitat, body size, and life rhythms merits further
investigation.

5 Reproductive Strategy

Long life and slowed-down biological rhythms also bear some relation to
a species' reproductive strategy. Among species where life is lived fast
and dangerously, the reproductive strategy is "altricial"—large numbers
of very immature offspring are born. They mature rapidly, and the
period of parental care is comparatively brief. In the "precocial" strategy
at the other end of the spectrum, one single offspring is produced, and
the nurturing is intensive and prolonged. In *Homo sapiens* it is more
intensive and prolonged than in any other known primate species.

The effect of aquatic life on a mammalian reproductive strategy cannot be accurately assessed because the terrestrial forebears of most aquatic mammals are extinct. An interesting comparison can be made, however, in the case of the mustelids (weasel family) where several closely related species are extant. The stoat, wholly terrestrial, bears annual litters of up to 12 or 13 young; the semi-aquatic river otter produces around four; the totally aquatic sea otter limits itself to a single pup every second year, which it nurses and supervises assiduously for over a year.

This precocial tendency appears to be strongest in secure and stable conditions with adequate and nonseasonal food supply—the reverse of conditions on the savannah. B. C. C. Rudder, toward the end of his thesis *The Allometry of Primate Reproductive Parameters,* comments about Hardy's aquatic theory: "The hypothesis has certain features that make it more attractive than the 'savannah' hypothesis. . . . It seems improbable on ecological grounds that savannah conditions result in anything but a deterioration in the level 1 efficiency for primates. This would not be a characteristic feature of an evolutionary advance."

Neoteny theory to date as applied to *Homo sapiens* has confined itself to a descriptive account of the mechanism by means of which certain changes evolved. Sooner or later further questions must be asked, such as: "Why did these happen to one primate and none of the others?" "When and where did the retardation begin to take place?" "What type of environment is most likely to have favored it?" The answers could prove interesting.

6 Brain Size

Before moving on to deal with similar questions about the aquatic theory itself, there is one more anomaly about human evolution that deserves mention.

Among primate species in general there has been a steady adaptive increase in relative brain size throughout the whole period of their evolution. R. D. Martin, in his paper *Adaptation and body size in*

Primates, demonstrates that even among the more primitive of extant primates, the prosimians (lemurs and lorises), this increase is detectable when they are compared with early fossil Paleocene and Eocene primates.

He concludes: "It would seem that brain size and foramen magnum area have both increased by adaptive shifts in primate evolution. If the two phenomena are connected, as the evidence suggests, then it seems likely that there might have been a close association between increasing locomotor sophistication and increase in brain size in primate evolution."

As with previous parameters, the brain size of *Homo sapiens* does not conform to the steady line of development traceable in other primate species. It takes a great leap forward.

Man's brain size now deviates from the mammalian norm to an extent that is shared only by the bottlenosed dolphin.

If it is true, as Martin suggests, that there may be a close association between increasing locomotor sophistication and increasing brain size, it is not easy to imagine how a descent from the trees to the savannah would have carried this process any further. It would be a descent to a much simpler and more stereotyped type of locomotion than that required in the arboreal habitat our predecessors had left behind.

On the other hand, moving to an aquatic environment demands the acquisition of an entirely new locomotor repertoire by every land animal that makes the transition. Limb movements, which on land had been automatic and stylized, in water became the subject of trial and error, conscious control and adjustment. And locomotor functions constitute a significant proportion of total brain function. Konrad Lorenz has commented on the fact that aquatic animals have relatively larger braincases than their nearest terrestrial relatives. The sea otter, for example, has a relatively larger braincase than the stoat. This fact could be dismissed as merely a consequence of increased body size. But not all such comparisons can be explained in this way. For example, in the case of the talapoin, the river-swimming monkey of Gabon, the ratio of braincase size to body weight is significantly larger than the primate norm.

If, then, following the acquisition of the new three-dimensional locomotor repertoire of swimming and diving, the aquatic ape returned to the land and was subjected to the necessity of consciously acquiring and perfecting yet another unprecedented locomotor pattern—bipedalism—then conceivably this rapid succession of evolutionary shocks would have stimulated his brain to the threshold of the period of unprecedented development on which it subsequently embarked.

The new high rate of human brain expansion was then maintained fairly consistently so that human brain evolution was a progressive process and became increasingly obvious after the advent of *Australopithecus*. It should also be noted that the influence on brain expansion of increased locomotor sophistication would be continued and enhanced by the newly acquired sophistication of other motor mechanisms—of the lips and tongue in speaking and of the fingers with their increasing precision and dexterity in manipulating objects of all kinds.

Such an environmental U-turn, from terrestrial to aquatic and back again, would be unique among primates. But the mind of man is unique also. It can hardly be accounted for by an evolutionary history that differs in no essential particular from that of the savannah baboon.

XI
Where and When
It Happened

1 *Ramapithecus* and *Australopithecus*

The crucial gap between manlike ape and apelike man is now considered by primate paleontologists to be that between *Ramapithecus* and *Australopithecus*. The remains of *Ramapithecus* are dated around 9 million years ago, and those of *Australopithecus*, discovered in Ethiopia and Tanzania, date from about 3.7 million years ago. Skeletons of *Australopithecus* show that he was clearly bipedal, and this is confirmed by the existence of a set of fossil footprints of two of these hominids walking side by side.

Reconstructions of the appearance of *Ramapithecus* have been used to illustrate countless books and articles—the hairy face and limbs, the long arms dangling forward, the stooped and slouching posture, the long toes and receding brow.

To date, all that has in fact been dug up of *Ramapithecus* consists of teeth and fragments of jawbone and palate. Everything apart from this is imaginative reconstruction, based on well-informed guesswork. There are no limb bones, no pelvises, no skulls (fig. 17).

This is not necessarily to cast doubt on the accuracy of the well-informed guesswork. But it is well to be reminded that conclusions based

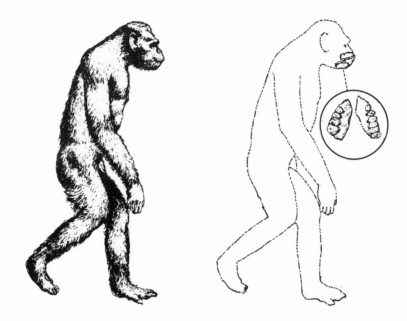

Fig. 17. Everything else is imaginative reconstruction of well-informed
          guesswork.

on fossil relics are neither more nor less "scientific" than conclusions
based on physiology. Anthropologists know that all other animals with
teeth and palate of that type also share other characteristics. The argument that all other animals with hairless skin and subcutaneous fat are,
or have been, aquatic is of precisely equal validity.

Relics of *Australopithecus* are much more plentiful and less fragmen-
tary. (One of them, Lucy, is nearly 40 percent complete.) However, the
fact that the fossils were more complete did not lead to more complete
agreement about their significance among the archeologists who discovered them. Mary Leakey argued that the fossils belong to our own
genus, *Homo*. Tim White and Donald Johanson believe they are bipedal
apes.

One result of the most recent discoveries has been further to discredit
the theory that bipedalism evolved to facilitate hunting and the carrying
of weapons.

Owen Lovejoy, writing about theorists who have "viewed hominization as the direct result of savannah occupation," remarks in a paper *The Origin of Man*: "There are many problems with this view. Bipedalism is useless for avoidance to escape from predators. Occasional bipedality, as seen in many primates, is sufficient for the use of weapons. More importantly, brain expansion and cultural development remotely postdate hominid divergence. . . . It is by now clear that man probably remained an omnivore throughout the Pleistocene, and that hunting may always have been an auxiliary food source."

Mary Leakey found strong evidence that no tools—or, at least, no durable tools—were being utilized by the *Australopithecines* who left their footprints in the Laetolil Beds: "We have encountered one anomaly. Despite three years of painstaking search by Peter Jones, no stone tools have been found in the Laetolil Beds. With their hands free, one would have expected this species to have developed tools or weapons of some kind. But, except for the ejecta of erupting volcanoes, we have not found a single stone introduced into the beds. We can only conclude, at least for the moment, that the hominids we discovered had not yet attained the tool-making stage."

2 Protein Dating

The fossil evidence leaves a gap of between five and six million years in which hominid divergence may have been initiated.

But fossils are not the only tools of the trade. Molecular biologists, such as Vincent Sarich and Allan Wilson of the University of California, claim that they can measure the degree of kinship between various species—and thereby deduce the date at which they first became distinct from one another—by "protein dating."

Protein dating is based on a study of the similarities and differences in the protein of living species. A. Zihlman and J. M. Lowenstein explain how it works:

"Proteins are made up of various combinations of the basic 20 amino acids, arranged in definite sequences. A given protein may include hundreds of thousands of amino acids. The proteins of closely related species, such as horse and donkey or dog and fox, are nearly identical, whereas species that diverged more than 100 million years ago, such as shrew and opossum, have many sequence differences. These differences can be measured precisely, and their number is approximately proportional to the divergence time."

If we accept the estimates of the molecular biologists, then the picture looks very different. They date the divergence between apes and hominids as being much more recent than previously supposed. (see graph, fig. 18.)

Not all scientists accept these figures. The protein-dating method is not 100 percent accurate. What it establishes is not an actual date, but the branching pattern for the tree linking various organisms, together with

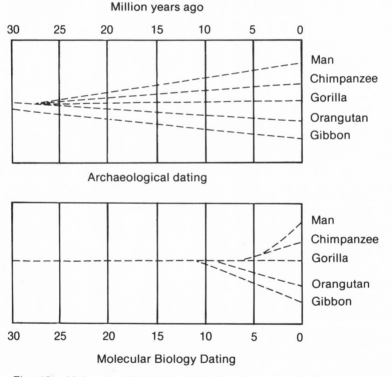

Fig. 18.   Molecular biologists claim the divergence between apes and hominids is much more recent.

some indication of the degree of change along various branches. It is based on the assumption that the rates of change of proteins during evolution remain uniform. This method has therefore been described as a "sloppy clock."

(The carbon-dating clock for determining the age of fossil relics was also rather sloppy, until refined by the use of dendrochronology—an ingenious method of dating based on tree rings.)

Nevertheless, when every allowance is made for sloppiness, we can no longer accept so confidently the hypothesis that apes evolved into men very gradually, becoming fractionally more upright and bipedal with every million years that elapsed.

It begins to look as if the thing happened, after all, rather more rapidly—a revolution rather than an evolution. And the more rapidly it took place, the more likely it is that some drastic changes in habitat or behavior are necessary to account for it. As long as conditions are stable, a species may continue unchanged for as long as the coelocanth. When conditions change very rapidly, then evolution goes into overdrive.

3 The Sea Came In. . . .

In parts of Africa, at the relevant period, precisely such drastic environmental changes were taking place. The sea came in and flooded vast areas in the north of the continent. Parts of the forested areas were cut off from the rest of Africa forming islands and sea marshes.

Populations of apes marooned on such islands may have found their usual food resources dwindling and turned to the sea that surrounded them for means of augmenting their diet. Some may have found their arboreal habitat turning into swampland and been forced to take to the water. Their incentive for learning to swim could have been similar to that of the proboscis monkeys in the often water-logged forests of Borneo.

It is possible that some such emergency, in much earlier times, accounted for the change to marine life of the ancestors of the whales, the seals, and the manatees, for, as J. Z. Young has pointed out, these species were originally quite as "ill-adapted" for such aquatic adventure as the African apes.

4 . . . And Went Out Again

If this hypothesis is correct, then *Australopithecus* was the ape that returned to the land.

Possibly he returned voluntarily, when the island or islands where he had been originally marooned became reunited with the African mainland. He would have followed the rivers and lakes upstream, southward, into the interior, maintaining a partly aquatic life style. Ultimately, the greater abundance of both game and vegetation would have induced him to spend more time on land and less in the water: the balance of advantage would then lie in a return to a terrestrial habitat.

Or conceivably his return to the land was forced on him, just as his original move into the sea had been forced on him. When the land rose and the sea retreated, vast pockets of sea water were cut off from the ocean, forming salt-water lakes that very gradually evaporated—as the Dead Sea is doing today, leaving deep layers of salt to mark where they had been. As the salinity increased they would have become uninhabitable, and the aquatic ape would have had no option but to return to life on land.

Whether the return was voluntary or involuntary, he would have returned greatly changed. The fossil record confirms that whatever happened to him during the five million year gap of which we have no record, he undeniably *did* return greatly changed. The Lucy fossil confirms that she was no longer a shambling stooping figure, but fully upright. The aquatic hypothesis would postulate that she was no longer a hairy ape-woman either, but as smooth skinned as her present day descendants.

Once nakedness had become the badge of her tribe, sexual selection may have favored its retention. Sexual selection frequently operates to reinforce differences between closely related species: the mate most likely to win favor is the one that displays in the highest degree those features that are most characteristic of the species and differentiate it from others. In the case of human beings hairlessness is such a feature. In the early post-aquatic days it would have been the most immediately striking difference, in visual terms, between ourselves and other anthropoids.

The return to terrestrial life meant a return to a habitat where olfactory signals—especially those governing sexual relationships—were normal and practicable; but their resuscitation would be hampered by the relatively atrophied state of human scent perception. This could' well explain the role of the axillary and urogenital scent organs which, according to W. Montagna (1975), are unique among the mammals. These "organs" consist of scent-secreting eccrine and apocrine glands in the underarm and genital areas, plus hair, which possibly served to retain the secretions and to facilitate the dispersal of the scent into the surrounding air.

The increase in braincase size, initiated by the U-turn but still comparatively unremarkable in *Australopithecus,* would thenceforward have been accelerated partly by the acquisition of speech, for which he was now strongly preadapted, and partly by the rapid increase in manual dexterity.

Alister Hardy (Appendix 2) has suggested some ways in which an aquatic habitat would have favored this latter development. The sea otter in a similar environment is an adroit tool user, employing round stones as hammers to loosen abalones from their undersea anchorage, and flat ones as anvils on which to smash open hard-shelled mollusks.

It has sometimes been argued that the aquatic theory is invalidated by the fact that no shell middens have been discovered. But there is no reason why the aquatic ape, any more than the sea otter, should have left shell middens: apes are not in the habit of collecting food and returning with it to a base or lair.

It seems probable that fish from the lakes and rivers continued to be part of the hominid's diet long after he had returned to a primarily terrestrial existence. Richard Leakey's 1981 book and television series *The Making of Mankind* featured a 1½-million-year-old skeleton of *Homo erectus* discovered on the shore of a lake in North Kenya. A team of pathologists, called in to diagnose a bone disease from which the hominid had suffered, identified the condition as "hypervitaminosis A"—a poisonous overdose of Vitamin A.

This was attributed by the team to an excessive intake of raw liver. But a toxic dose of the vitamin would necessitate a large daily intake of liver over a period of months—hardly a typical diet for any hunting/gathering or scavenging group. By far the commonest source of Vitamin A, as one of the doctors pointed out, is fish. The suggestion he made—that this particular lake dweller was a "fish eater"—seems the likeliest explanation.

Meat eating itself could well have begun on the seashore. Some of the animals the ape would have encountered in the shallows or on the beaches (sea turtles, dugongs) were large, docile, and helpless on land. Their presence would have encouraged him to begin thinking of himself as a predator, a role for which he was not naturally well equipped. He had no natural weapons such as the claws of the big cats or the kind of teeth that would easily penetrate the hides of larger prey. The consequent necessity for skinning and dismembering his victims would have provided further incentive to develop tool-using and tool-making techniques.

5 The Geological Background

The idea that the inundation of parts of northern Africa formed the background for this aquatic interlude was put forward by Leon P. La Lumiere, Jr., of the Naval Research Laboratory in Washington. (See Appendix 1.)

Starting from Hardy's original hypothesis, La Lumiere reasoned that if this aquatic interlude took place at all it must have had as its starting point a forested area inhabited by apes in the late Miocene. He then postulated that speciation of the kind indicated by the fossil record strongly suggested that a population of the apes had become isolated over a prolonged period from others of their kind; and that any such isolation must have come to an end in the late Pliocene or early Pleistocene to account for the siting of early hominid remains.

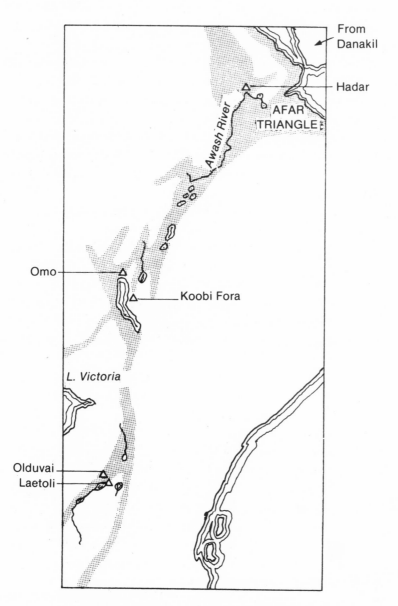

They followed the water-courses south . . .

Fig. 19. The African Rift Valley would have been the obvious route of migration followed by the aquatic ape.

Consequently, La Lumiere embarked on a detailed study of the geological history of Africa, in order to ascertain whether there was any area of the continent that fulfilled all these conditions ecologically, geographically and chronologically. He discovered such an area in the northern and central Afar triangle, specifically the region of the Danakil Alps, which was at the relevant period effectively cut off from the rest of Africa by the sea.

He provides a fully documented account of the geological events contemporary with the period of man's emergence, and the reasons why he believes that, when the Danakil region was rejoined to the mainland of Africa, the likeliest route along which the hominids would have migrated would have been southward along the African Rift Valley (see fig. 19).

The evolutionary story being unearthed by succeeding fossil discoveries is a shadowy and ever-changing one. Stephen Jay Gould vividly underlined this fact when he wrote: "New and significant prehuman fossils have been unearthed with such unrelenting frequency in recent years that the fate of any lecture notes can only be described with the watchword of a fundamentally irrational economy—planned obsolescence." It may be that some of the minor details of La Lumiere's hypothesis will prove to be subject to the same fate and need further modification, but in its broad outline it must be regarded as a durable and genuinely illuminating contribution which adds a new dimension to the debate on man's origins.

It is persuasive on many grounds:

(1) The geological facts about the inundation are not challenged. La Lumiere's contribution was to perceive their relevance to the aquatic theory. Populations of apes must undoubtedly in those circumstances have found themselves isolated on Danakil, or on smaller islands, or marooned in the treetops of lower-lying forest areas invaded by the sea. Some, possibly a high percentage in the affected areas, would have succumbed to such a drastic change in their environment. But the primates are a relatively adaptive order. Some individuals would have survived and adapted to the changed conditions. This was the line that led to Man.

(2) The timing of the geological events outlined is not incompatible with either of the currently proposed time scales of human evolution—that of the paleontological archaeologists, or that of the molecular biologists.

(3) It provides a possible answer to one of the most puzzling aspects of the aquatic theory—namely, if the aquatic ape was adjusting successfully to some kind of aquatic life, why then did he subsequently leave the sea? La Lumiere's scenario opens the possibility that up to a point the sea left the aquatic ape. This then redirected the stranded ape's attention to the opportunities on land, which he was now in several ways better equipped to exploit.

(4) The conviction is growing among evolutionists that the emergence of man was accomplished more rapidly than was once imagined. La Lumiere postulates that the process was initiated on an island; and islands have always been forcing-houses of evolutionary change and speciation, as Darwin's Galapagos observations made clear.

(5) The maps (page 126) show the high probability that the African Rift Valley would have been the obvious route of migration followed by an ex-aquatic ape. Most of the significant discoveries of African prehuman fossil remains have been found in this area (see fig. 20:3).

Professor D. R. Newth's comment on the aquatic theory (*Nature*, April, 1982) was: "The fossil record will doubtless have the final say."

The five-million year gap in that fossil record means that until more fossils come to light *all* hypotheses must remain speculative. The aquatic theory would predicate that if bones are found dating from the period of the fossil gap they will be found not far from the Afar triangle.

It would also predicate that changes in the alignment of the bones in the pelvic area and in the position of the foramen magnum (indicating the angle of the head) will be found to have *preceded* the changes in the bones of the knees, feet, and ankles necessitated by the weight-bearing stresses of bipedal locomotion. In other words, I believe that the hominid straightened out before he habitually stood up and walked. If this is true, the fossil finds will one day prove it—just as the Lucy skeleton proved that he stood upright and walked before his brain got bigger.

Pending such discoveries, we can only argue from probabilities, and there are three factors that render the aquatic theory a tenable one:

(1) It accounts for more of the differences between man and other primates than does any alternative theory.

(2) As the fossil gap shortens, it becomes harder to believe that one group of apes would have evolved so rapidly along lines so radically different from all other groups, if it had continued to share with them the same unchanging environment.

(3) There is no incompatibility between the aquatic theory and the other ideas that have been advanced to account for the emergence of man. The aquatic interlude was the trigger of change; neoteny was the mechanism of change; and the savannah was the domain that the aquatic ape returned to conquer.

Appendix 1
Danakil Island

THE EVOLUTION OF HUMAN BIPEDALISM:
where it happened—a new hypothesis.
by

Leon P. La Lumiere, Jr.
(Naval Research Laboratory,
Washington, D.C., U.S.A.)

A geological plausible locality where aquatic evolutionary processes leading to bipedalism could have occurred is postulated. A time period for these processes is suggested, filling an hiatus in current theory. A geological formation likely to contain fossils is identified. Explanations for contemporaneity of various early fossil hominids are suggested.

Introduction

(After outlining the aquatic theory, La Lumiere proceeds:)

So far, the aquatic hypothesis has received little acceptance because no supporting fossil evidence has been adduced (Morris, 1967; Young, 1971). In particular, no region in Africa containing marine Pliocene

123

deposits associated with apelike and manlike fossils has ever been found (Howells, 1967; Leakey, 1976).

Reflection upon the Hardy hypothesis leads to the following conclusions: (a) the region was a forested area inhabited by apes during the late Miocene; (b) the region was isolated from the rest of Africa during the Pliocene in which period the evolution of apelike to manlike creatures occurred; (c) the region was reconnected to Africa in the late Pliocene or early Pleistocene enabling the hominids to migrate elsewhere. Part (b) suggests an island.

Evidence for the Hypothetical Locality

Tazieff (1972), Tazieff *et al* (1972), and Barberi *et al* (1972) suggest that the northern and central Afar triangle in the past was covered by seawater with only the Danakil Alps and high volcanoes standing above water as islands. They state that the Danakil Alps are part of a horst: an uplifted crustal block that was broken off and separated from the Nubian plate to the west and the Arabian plate to the east through the action of plate tectonics and seafloor spreading.

The paradigm of continental drift and seafloor spreading described by Wegener (1966), Wilson (1963), Bullard (1972), and Dietz and Holden (1972) along with others provides insight and understanding of the geophysical evolution of the Afar triangle. Near the end of the Oligocene or the beginning of the Miocene, the African plate, of which the Arabian plate was then a part, apparently collided with the Eurasian plate. This collision may have caused three tectonic events that are germane to this paper, namely:

(1) Doming uplifted the region perpendicular to the axis of the present day Red Sea causing cracks in the Afro-Arabian plate, and was followed by downfaulting and the formation of the proto-Red Sea, which was connected with the proto-Mediterranean Sea (Hutchison & Engels 1970, 1972; Coleman 1974; Pilger & Rosler 1976).

(2) Rifting began in the Gulf of Aden (Hutchison & Engels 1970, 1972; Coleman 1974; Pilger & Rosler 1976).

(3) At the juncture of these two regions tectonic activity produced crustal blocks of assorted sizes (Tazieff 1972; Tazieff *et al* 1972).

According to Hsu, Ryan, and Cita (1973) starting at the beginning of the Messinian (latest Miocene stage) the Mediterranean Sea was repeatedly isolated and then rejoined to the Atlantic Ocean causing the sea to dry up and then refill. They suggest this cycle of drying and refilling was repeated at least eleven times and perhaps as many as fourteen. During the desiccation of the sea, massive thicknesses of salt were deposited on the bottom of the deeper parts.

After combining the results of biostratigraphic, chronostratigraphic, and paleomagnetic investigations Cita and Ryan (1973) devised an absolute time scale which they believe is reliable and useful. They suggest the Miocene-Pliocene boundary should be set at about 5.4 million years before present (mybp). They also suggest the Messinian began shortly after 7.5 mybp.

According to Coleman (1974) the Red Sea was a gulf connected to the Mediterranean Sea during the Late Miocene. Stoffers and Ross (1974) suggest that when the Mediterranean was subjected to the cycle of drying and refilling, so also was the Red Sea. During this period, massive thicknesses of salt were deposited on the bottom of the deeper parts of the sea.

Figure 20:1 displays the configuration of the African continent and Arabian plate as it may have been during the late Miocene. It should be noted that the proto-Red Sea and the proto-Gulf of Aden were separated by an isthmus. This land bridge, here called the Afar Isthmus, apparently existed throughout the late Miocene and was an important link in animal migration between the continents of Africa and Eurasia (Kurten 1972; Beyth 1977; MacKinnon 1978).

Many anthropologists suggest that *Ramapithecus* was an ancestor of the genus *Homo* (Howells 1967; Young 1971; Leakey 1976; Leakey and Lewin 1977, 1978; MacKinnon 1978). Fossils of this ape from Fort Ternan, Kenya, have been dated at 12.5-14 mybp and from the Siwalik Hills in India at 9-12 mybp (Leakey 1976). Regardless of their origin these apes must have migrated between Africa and Asia across the

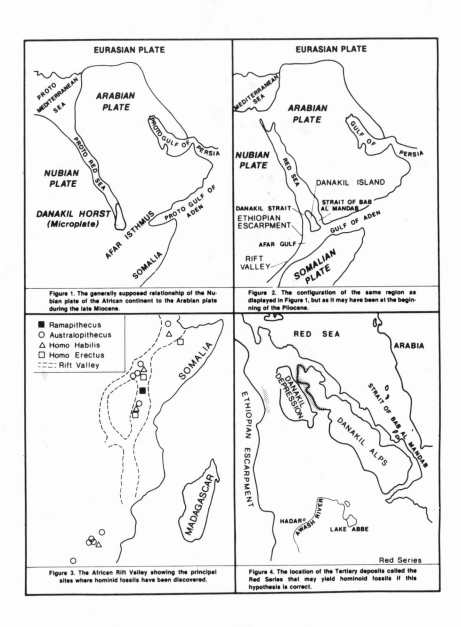

Figure 1. The generally supposed relationship of the Nubian plate of the African continent to the Arabian plate during the late Miocene.

Figure 2. The configuration of the same region as displayed in Figure 1, but as it may have been at the beginning of the Pliocene.

Figure 3. The African Rift Valley showing the principal sites where hominid fossils have been discovered.

Figure 4. The location of the Tertiary deposits called the Red Series that may yield hominoid fossils if this hypothesis is correct.

Fig. 20.   La Lumiere's maps

isthmus connecting the two continents. Furthermore, as the apes spread out, groups would occupy and live in favorable niches such as the forests of the isthmus.

The Afar Isthmus was composed of several crustal blocks. One of these, the Danakil horst, apparently acted as a "microplate" (Le Pichon & Francheteau, 1978). The horst is a mountainous region about 335 miles (540 km) long and up to 45 miles (75 km) wide (Tazieff *et al* 1972; Geol. Surv. 1973). In Figure 1, the northern end of the horst marks the southern limit of the proto-Red Sea (Frazier 1970; Barberi *et al* 1972). It is, therefore, reasonable to suppose that the horst was occupied about 9-14 mybp by a group of apes who continued to live there until the forests disappeared at the end of the Miocene (Kurten 1972; MacKinnon 1978).

Volcanism within the Afar triangle and adjacent regions was a direct result of plate rupture and rifting. However, volcanism was episodic: periods of intense activity were followed by periods of quiescence (Gass 1974; Pilger & Rosler 1976). Barberi *et al* (1972) have radiometrically dated samples of volcanic rocks from various localities in the Afar to provide a brief geo-chronological history of the region. According to them, the oldest volcanic rocks are about 25 million years old. In the early Miocene, the Danakil horst was separated from the Ethiopian escarpment to form a depression extending southward from the north end of the horst about one-third to one-half its length (Hutchison & Engels 1970, 1972; Barberi *et al* 1971). Within the Danakil depression, as it is called, from its inception to the present time, volcanism and sedimentation of marine, lacustrine, evaporitic, and continental focies have occurred contemporaneously (Barberi *et al*). The depression apparently was a lake or embayment that was responsible for the deposition of detrital formation known as the Red Series. These deposits have an age range of 5-24 mybp (Barberi *et al* 1972) and currently exist as narrow bands on each side of the Danakil depression (Geol. Surv. 1973). Much of the Red Series is overlain by Quarternary deposits or lava flows. According to Barberi *et al* (1972) the age of the upper part of the Red Series is 5.4 mybp. Starting in the early Miocene, one or more volcanoes in the middle of the horst erupted intermittently, producing extensive lava flows that still cover the entire southern portion (Barberi *et al* 1972). Lava also covers the far northern end. In between lies a region about 90 miles (150 km) long and about 45 miles (75 km) wide in which the exposed formations are Mesozoic rocks, mainly Jurassic, covered by a discontinuous veneer of Tertiary and Quaternary deposits (Hutchison & Engels 1972).

After a long period of relative quiescence, starting about 9-11 mybp volcanism increased in the Red Sea, in the Afar triangle, and in the African rift valley indicating renewed tectonic activity (Pilger & Rosler 1976). About 6.7 mybp the Danakil depression was invaded by marine waters (Barberi *et al* 1972). Hutchison and Engels (1972) state that the deeper part of the depression is covered by thick salt deposits. They suggest that the deposition of the lower layers may be correlative with the upper Miocene salt deposits of the Red Sea. The cycle of drying and refilling that apparently occurred in the Mediterranean Sea and in the Red Sea also may have occurred in the Danakil depression.

About the Miocene-Pliocene boundary the African plate moved away from the Arabian plate, and the Danakil microplate was rotated counter-clockwise (Tazieff *et al* 1972; Le Pichon & Francheteau 1978). At the same time it was tilted so that its Mesozoic sedimentary rock formations slope generally from northeast to southwest (Hutchison & Engels 1970, 1972; Beyth 1978). Excluding volcanic peaks, the Danakil Alps today rise to a maximum of 4,455 feet (1,335 m). Finally, the microplate was detached from both the African and Arabian plates, allowing waters from the Red Sea and the Gulf of Aden to flow into the Afar triangle. Figure 20:2 displays the configuration of the region as it may have been at the beginning of the Pliocene. Note that the Red Sea was no longer connected to the Mediterranean Sea (Coleman 1974) as in figure 20:1, but was now linked to the Gulf of Aden and the Indian Ocean through two straits, one to the east of the Danakil horst (Strait of Bab al Mandab) and the other to the west that will be called the Danakil Strait. Thus, between 6.7 and 5.4 mybp in the latest Miocene (Messinian), a group of apes along with other animals could have been trapped on Danakil Island.

According to Barberi *et al* (1972) and Mohr (1978) the central and southern Afar regions have been repeatedly covered by massive flood basalts during the Pleistocene-Holocene so that the Miocene-Pliocene history of these regions is uncertain. The several volcanoes in the middle of the Danakil horst have been intermittently active from the late Miocene-early Pliocene to the present. About the Miocene-Pliocene boundary the Danakil horst apparently was surrounded by water to the east, the north, and the west while the southern end was covered by extensive flood basalts. The Danakil horst may not have been a geographic island, but under the conditions described above, for many land animals, it would have been a biological island.

A Scenario for the Evolution of Genus Homo

The following is a tentative hypothesis describing what may have happened. Although speculative, it is well to show that some such course of events may be envisaged.

The Pliocene was a time of increasing desiccation. (If the hypothesis of Hsu, Cita, and Ryan (1973) is correct, desiccation may have started during the Messinian.) Forests probably covered most of Danakil Island at the beginning of the Pliocene, but these must have soon died. Those near sea level and the coast would be the first to disappear, while those at higher, cooler elevations in the mountains remained longer. The dwindling forest would produce exactly the environmental conditions required by the Hardy hypothesis: those apes near the coast, losing their forest, gradually would be forced into water to find both food and protection from predators. Increasing dryness would likely destroy much vegetation, reducing the population of both herbivores and carnivores.

Those apes living along the coast, would likely wander back and forth searching for food by wading in the shallow waters: a behavior often compelling upright bipedal movement. (Their cousins along the Ethiopian escarpment and elsewhere undoubtedly retreated with the dwindling forests.) Thus, the island coastal apes, forced to live under unusual conditions, would rapidly evolve into the upright, hairless creatures described by Hardy and Morgan. . . .

... Witnessing frequent volcanic eruptions and lava flows at both the north and south ends of the island, the apes may have made two important discoveries: pebble tools and fire. Hot lava passing over pebbles scattered along the beaches could have shattered these, to produce keen-edged shards; on meeting water, lavas would be cooled and sundered into sharpened ready-made tools. Lavas may have cooked plants and animals and so led the apes to consume and appreciate cooked food: a knowledge that would become invaluable to their descendants in the African rift valley.

Sporadic and episodic volcanism within the Afar triangle has been a feature since the early Miocene (Barberi *et al* 1972; Gass 1974). Intermittently, the Danakil Strait has been closed and bridged by lava flows as it is today (Frazier 1970; Hutchison & Engels 1972; Lowell & Genik 1971). Eustatic sea level fluctuations combined with erosion probably reopened the strait within a short time. However, during the short time the island was connected to the mainland, migration of animals must have occurred with the *Australopithecines* among them.

These hominids had evolved in and near the water, and as they wandered over the lava bridge to the Ethiopian escarpment and then elsewhere, they stayed near the water. They did so for two reasons. First, water was their protection against predators. Next, water provided them with food and drink.

In the meandering search for food, the hominids drifted southward along the western shores of the Afar Gulf. Whenever possible, they explored the rivers and streams that emptied into the embayment. About three million years ago, some of the hominids settled in a place now called Hadar, near the Awash River (Johanson & White 1979).

The African rift valley is a consequence of the collision and subsequent separation of the African and Eurasian plates. Figure 20:3 displays its location. Since its inception, lakes and rivers have been created. Many have been filled to cover and to preserve fossils. In particular, the Awash River emptied into the Afar Gulf. The river course follows the rift valley in a generally southwestern direction from the Afar. The hominids followed the river upstream, eventually arriving at the Omo valley and then proceeded southward leaving their remains along the way at such present-day sites as the Omo River, Koobi Fora, Lake Turkana, Olduvai Gorge, Laetolil, Makapansgat, Sterkfontein, and Taung (Leakey & Lewin 1978). These sites are located in Figure 3. The oldest known hominid fossil is about 3.8 million years old (Leakey 1976; Johanson & White 1979), with those found at Hadar being more primitive than those from South Africa.

Biochemical studies indicate that the hominid line of evolution diverged from the apes and gorillas at least 5 mybp (Sarich & Wilson 1967) and perhaps as much as 10-13 mybp (Benveniste & Todaro 1976; MacKinnon 1978).

The foregoing suggests that the aquatically-evolving apes were isolated on Danakil Island for at least one and one-half million years and perhaps as long as three million years before returning to the mainstream of African life.

Although some *Australopithecines* left the island, others remained behind to continue their evolution. About 1.75 mybp, a group of hominids crossed the land bridge, followed the rift valley into the interior and left remains at Koobi Fora and Olduvai Gorge. These hominids are now called *Homo habilis* (Leakey 1976; Leakey & Lewin 1977, 1978).

Following others, Morgan remarks that an aquatic environment produces greater pressure to evolve than does a terrestrial one. This may explain why *Homo babilis* appeared in the rift valley while both types of *Australopithecines* were still extant.

As *Homo babilis* gradually replaced the *Australopithecines,* others of his kind remained behind on the shores of Danakil Island to continue their evolution. These descendants of *Homo habilis* left the island when opportunity permitted about one million years ago. They are now known as *Homo erectus*. Since the Suez Isthmus had emerged at the beginning of the Pliocene (Coleman 1974), they were able to spread throughout the eastern hemisphere (Kurten 1972; MacKinnon 1978).

A pattern of isolation, evolution, and escape of some hominid inhabitants of Danakil Island appears likely. Starting some five million or more years ago, this cycle may have been repeated until about 30,000 years ago when the final desiccation of the Danakil Depression occurred (Bonatti *et al* 1971; Hutchison & Engels 1972; Tazieff 1972).

Suggested Locality for Exploration

Tectonic activity is extremely high within the Afar triangle (Tazieff 1972). Much of the region is covered by flood or plateau basalts (Geol. Surv. 1973), and exposed continental basement is limited; quaternary and recent deposits are more extensive and cover both basement and basalts, especially along the Ethiopian and Somalian escarpments and the coast between the Danakil Alps and the Red Sea. The Red Series are Tertiary deposits that contain Miocene fossils (Frazier 1970; Hutchison & Engels 1970; Beyth 1978). The radiometric age of the series ranges from 5.4 to 24.0 mybp (Barberi *et al* 1972). Deposits occur along the foothills of the western edge of the northern section of the Danakil Alps and east of the Danakil Depression. They also occur along the foothills of the Ethiopian escarpment west of the Depression. The location of the Red Series is indicated in figure 20:4. If this hypothesis is correct, fossils of our apelike and manlike ancestors should be found in them and in Quaternary formations.

Conclusions

This hypothesis combined with those of Hardy and Morgan suggests some answers, implicitly as well as explicitly, to many questions posed by students of human evolution. The main points are:

(1) A geologically plausible locality where aquatic evolutionary processes could have occurred is postulated.
(2) A time zone for the processes is suggested, filling an hiatus in current theory.
(3) A geologic formation likely to contain hominoid as well as hominid fossils is identified.
(4) Explanations for contemporaneity of various early fossil hominids is suggested.

Acknowledgments

The author extends his thanks for assistance with articles and manu-
scripts, comments, criticisms, and suggestions to N. Z. Cherkis, R. H.
Feden, H. F. Fleming, R. K. Perry, P. R. Vogt, and A. C. Hardy. He
thanks E. J. Andersen and C. S. Fruik for drawing the accompanying
figures and J. F. Peery, R. H. Baturin, J. A. Marshall, and D. Love for
typing the manuscript.

References

Barberi, F., Borsi, S., Ferrara, G., Marinelli, G., Santacroce, R., Tazieff, H.,
  & Varet, J. 1972 *J. Geol.* 80, 720-29.
Benveniste, R. E., & Todaro, G. J. 1976 *Nature* 261, 101-08.
Beyth, M. 1978 *Tectonophysics* 46, 357-67.
Bonatti, E., Emiliani, C., Ostlund, G., & Rydell, H. 1971 *Science* 172,
  468-69.
Bullard, E. 1972 The Origin of the Oceans; in *Continents Adrift*, San
  Francisco: W. H. Freeman & Co., 88-97.
Cita, M. B. & Ryan, W. B. F. 1973 Time Scale and General Synthesis; in
  Ryan, W. B. F., Hsu, K. J. et al (eds) 1973 *Initial Reports of the
  Deep Sea Drilling Project*, Volume XIII, Part 2, Washington: U.S.
  Government Printing Office, 1405-15.
Coleman, R. G. 1974 Geologic Background of the Red Sea; in Burk, C. A.
  & Drake, C. L. (eds) *The Geology of Continental Margins*, New
  York: Springer-Verlag, 743-51.
Dietz, R. S. & Holden, J. C. 1972 The Breakup of Pangaea; in *Continents
  Adrift*, San Francisco: W. H. Freeman & Co., 102-13
Frazier, S. B. 1970 *Phil. Trans. Roy. Soc. Lond.* A. 267, 131-41.
Gass, I. 1974 *Nature* 249, 309-10.
Geological Map of Ethiopia 1973, 1/2000000, Geol. Surv. of Ethiopia.
Hardy, A. C. 1960 *The New Scientist* 7, 642-45.
Howells, W. W. 1967 *Mankind in the Making*, Garden City, N.Y.:
  Doubleday.

Hsu, K. J., Cita, M. B., & Ryan, W. B. F. 1973 The Origin of the Mediterranean Evaporites; in Ryan, W. B. F., Hsu, K. J. et al (eds) 1973 *Initial Reports of the Deep Sea Drilling Project,* Volume XIII, Part 2, Washington: U.S. Government Printing Office, 1203-31.

Hutchison, R. W. & Engels, G. G. 1970 *Phil. Trans. Roy. Soc. Lond.* A. 267, 313-29.

Hutchison, R. W. & Engels, G. G. 1972 *Geol. Soc. Am. Bull.* 83, 2989-3002.

Johanson, D. C. & White, T. D. 1979 *Science* 203, 321-30.

Kurten, B. 1972 *The Age of Mammals,* New York: Columbia University Press.

Leakey, R. E. F. 1976 *American Scientist* 64, 174-78.

Leakey, R. E. F. & Lewin, R. 1977 *Origins,* New York: E. P. Dutton.

Leakey, R. E. F. & Lewin, R. 1978 *People of the Lake,* Garden City, N.Y.: Anchor Press, Doubleday.

Le Pichon, X. & Francheteau, J. 1978 *Tectonophysics* 46, 369-406.

Lowell, J. D. & Genik, G. J. 1971 *Bull. Am. Assoc. Petrol. Geo.* 56, 247-59.

MacKinnon, J. 1978 *The Ape Within Us,* New York: Holt, Rinehart & Winston.

Mohr, P. A. 1978 *Ann. Rev. Earth. Planet. Sci.* 6, 145-72.

Morgan, E. 1972 *The Descent of Woman,* New York: Stein & Day.

Morris, D. 1967 *The Naked Ape,* New York: McGraw-Hill.

Morris, D. 1977 *Manwatching,* New York: Abrams.

Pilger, A. & Rosler, A. 1976 *Abhandlungen der Braunschweigishen Wissenschaftlichen Gesellschaft* 26, 67-90.

Sarich, V. M. & Wilson, A. C. 1967 *Science* 158, 1200-03.

Stoffers, P. & Ross, D. A. 1974 Sedimentary History of the Red Sea; in Whitmarsh, R. B., Weser, O. E., Ross, D. A. et al (eds) 1974 *Initial Reports of the Deep Sea Drilling Project,* Volume XXIII, Washington: U.S. Government Printing Office, 849-65.

Tazieff, H. 1972 The Afar Triangle; in *Continents Adrift,* San Francisco: W. H. Freeman & Co., 133-41.

Tazieff, H., Varet, J., Barberi, F., & Giglia, F. 1972 *(Nature* 235, 144-47.

Wegener, A. 1966 *The Origin of Continents and Oceans* (Translated from the fourth revised German edition by John Biram, New York: Dover Publications).

Wilson, J. T. 1963 *Nature* 198, 925-29.

Young. J. Z. 1971 *An Introduction to the Study of Man,* Oxford at the Clarendon Press.

The contents of the above paper were communicated by Sir Alister Hardy to the Royal Society during a symposium on the Emergence of Man and subsequently published in the Society's Philosophical Transactions (Phil. Trans. R. Soc., London, B 292, 103-07 (1981).

Appendix 2

Sir Alister Hardy's aquatic hypothesis stated in the words of the only three accounts he has so far published.

Statement 1

This is the account given in *The New Scientist,* Vol. 7, pages 642-45, April, 1960. It appeared as follows:

> WAS MAN MORE AQUATIC IN THE PAST?
> And was it in the sea that man learned to stand erect? The author explains his hypothesis that we descend from more aquatic ape-like ancestors.
> By Professor SIR ALISTER HARDY, FRA.

On 5 March I was asked to address a conference of the British Sub-Aqua Club at Brighton and chose as my theme "Aquatic Man: Past, Present and Future." I dealt little with the present, for Man's recent achievements in the underwater world were so well illustrated by other speakers and by films. I ventured to suggest a new hypothesis of Man's origins from more aquatic apelike ancestors and then went on to discuss possible developments of the future. I did not expect the wide publicity that was given to my views in the daily press, and since such accounts could only be much abbreviated, and in some cases might be misleading, I gladly accepted the invitation of *The New Scientist* to give a fuller statement of my ideas.

I have been toying with this concept of Man's evolution for many years, but until this moment, which suddenly appeared to be an appropriate one, I had hesitated because it had seemed perhaps too fantastic, yet the more I reflected upon it, the more I came to believe it to be possible, or even likely. In this article I shall deal with this hypothesis; next week I shall treat of the future.*

Man, of course, is a mammal, and all the mammals have been derived, as indeed have also the birds but by a different line of evolution, from reptile ancestors that flourished more than a hundred million years ago, when the world was populated by saurians of so many different kinds which have long since become extinct. These reptile ancestors in turn were derived from newt-like animals—amphibian creatures—which had only partially conquered the land and had to return to water to breed as do most of our salamanders and frogs of today. It is equally certain that these earlier amphibians were evolved from fish which, like those primitive lung-fish that still survive in certain tropical swamps today, had developed lungs with which to breathe. Some of these air-breathing fish were able to climb from the water on to the land.

This history of the emancipation of animal life from the sea is well known. I repeat it only because it forms the background to another story, one that is not quite so familiar to those who are not trained as zoologists. At the same time as this conquest of the land was extending with continuously improving adaptations to the new terrestrial life, we see (in the fossil record) a different act repeating itself again and again, first with the amphibians, next with the reptiles, and then with the mammals and indeed the birds as well. Excessive multiplication, over-

---

*In the second article the author developed his ideas as to how sub-aqua man in the future would revolutionize the fishing industry.

population, shortage of food, resulted in some members of each group**
being forced back into the water to make a living, because there was not
enough food for them on the land. Among the reptiles I need only
remind you of the remarkable fish-like ichthyosaurs, of the plesiosaurs,
of many crocodile-like animals, and of turtles, not to mention water-
snakes. Then, among the mammals of today we see the great group of
whales, dolphins and porpoises, with the vestigial remains of their hind
legs buried deep in their bodies, beautifully adapted to sea life; or again
the dugongs and manatees belonging to an entirely different order. The
seals are well on their way to an almost completely aquatic life, and many
other groups of mammals have aquatic representatives which have been
forced into the water in search of food: the polar bears, the otters
(both freshwater and marine), various aquatic rodents, like water voles
and the coypus, or insectivores like the water shrew; and, of course, we
must not forget the primitive duck-billed platypus. There are, indeed,
few groups that have not, during one time or another in the course of
evolution, had their aquatic representatives; among the birds the penguins are the supreme examples.

The suggestion I am about to make may at first seem farfetched, yet I
think it may best explain the striking physical differences that separate
Man's immediate ancestors (the Hominidae) from the more ape-like
forms (Pongidae) which have each diverged from a common stock of
more primitive ape-like creatures which had clearly developed for a time
as tree-living forms.

My thesis is that a branch of this primitive ape-stock was forced by
competition from life in the trees to feed on the seashores and to hunt
for food, shell fish, sea-urchins, etc., in the shallow waters of the coast. I
suppose that they were forced into the water just as we have seen
happen in so many other groups of terrestrial animals. I am imagining
this happening in the warmer parts of the world, in the tropical seas
where Man could stand being in the water for relatively long periods,

**The amphibians went back only into fresh water (for certain physiological
reasons) *not* into the sea.

that is, several hours at a stretch. I imagine him wading, at first perhaps still crouching, almost on all fours, groping about in the water, digging for shell fish, but becoming gradually more adept at swimming. Then, in time, I see him becoming more and more of an aquatic animal going further out from the shore; I see him diving for shell fish, prising out worms, burrowing crabs and bivalves from the sands at the bottom of shallow seas, and breaking open sea-urchins, and then, with increasing skill, capturing fish with his hands.

Let us now consider a number of points which such a conception might explain. First and foremost, perhaps, is the exceptional ability of Man to swim, to swim like a frog, and his great endurance at it. The fact that some men can swim the English Channel (albeit with training), indeed that they race across it, indicates to my mind that there must have been a long period of natural selection improving Man's qualities for such feats. Many animals can swim at the surface, but few terrestrial mammals can rival man in swimming below the surface and gracefully turning this way and that in search of what he may be looking for. The extent to which sponge and pearl divers can hold their breath under water is perhaps another outcome of such past adaptation.

It may be objected that children have to be taught to swim; but the same is true of young otters, and I should regard them as more aquatic than Man has been. Further, I have been told that babies put into water before they have learnt to walk will, in fact, go through the motions of swimming at once, but not after they have walked.

Does the idea perhaps explain the satisfaction that so many people feel in going to the seaside, in bathing, and in indulging in various forms of aquatic sport? Does not the vogue of the aqua-lung indicate a latent urge in Man to swim below the surface?

Whilst not invariably so, the loss of hair is a characteristic of a number of aquatic mammals, for example, the whales, the Sirenia (that is, the dugongs and manatees) and the hippopotamus. Aquatic animals which come out of the water in cold and temperate climates have retained their fur for warmth on land, as have the seals, otters, beavers, etc. Man has lost his hair all except on the head, that part of him sticking out of the water as he swims; such hair is possibly retained as a guard against the rays of the tropical sun, and its loss from the face of the female is, of course, the result of sexual selection. Actually the apparent hairlessness of Man is not always due to an absence of hair; in the white races it is more apparent than real in that the hairs are there but are small and exceedingly reduced in thickness; in some of the black races, however, the hairs have actually gone, but in either case the effect is the same: that of reducing the resistance of the body in swimming. Hair, under water, naturally loses its original function of keeping the body warm by acting as a poor heat conductor; that quality, of course, depends upon the air held stationary in the spaces between the hairs—the principle adopted in Aertex underwear. Actually the loss or reduction of hair in Man is an adaptation by the retention into adult life of an early embryonic condition; the unborn chimpanzee has hair on its head like Man, but little on its body.

While discussing hair it is interesting to point out that what are called the "hair tracts"—the direction in which the hairs lie on different parts of the body—are different in Man from those in apes; particularly to be noted are the hairs on the back, which are all pointing in lines to meet diagonally toward the mid-line, exactly as the streams of water would pass around the body and meet, when it is swimming forward like a frog. Such an arrangement of hair, offering less resistance, may have been a first step in aquatic adaptation before its loss.

The graceful shape of Man—or woman!—is most striking when compared with the clumsy form of the ape. All the curves of the human body have the beauty of a well-designed boat. Man is indeed streamlined.

These sweeping curves of the body are helped by the development of fat below the skin and, indeed, the presence of this subcutaneous fat is a characteristic that distinguishes Man from the other primates. It was a note of this fact in the late Professor Wood Jones's book *Man's Place among the Mammals* (p. 309) that set me thinking of the possibility of Man having a more aquatic past when I read it more than thirty years ago. I quote the paragraph as follows:

"The peculiar relation of the skin to the underlying superficial fascia is a very real distinction, familiar enough to everyone who has repeatedly skinned both human subjects and any other members of the Primates.

The bed of subcutaneous fat adherent to the skin, so conspicuous in Man, is possibly related to his apparent hair reduction; though it is difficult to see why, if no other factor is invoked, there should be such a basal difference between Man and the Chimpanzee."

I read this in 1929 when I had recently returned from an Antarctic expedition where the layers of blubber of whales, seals, and penguins were such a feature of these examples of aquatic life; such layers of fat are found in other water animals as well; and at once I thought perhaps Man had been aquatic too. In warm-blooded water animals such layers of fat act as insulating layers to prevent heat loss; in fact, in function they replace the hair. Man, having lost his hair, must, before he acquired the use of clothing, have been subjected to great contrasts of temperature out of water; in this connection it is interesting to note the experiments carried out at Oxford by Dr. J. S. Weiner, who showed what an exceptional range of temperature change in air Man can stand, compared with other mammals. Man's great number of sweat glands enable him to stand a tropical climate and still retain a large layer of fat necessary for aquatic life.

This idea of an aquatic past might also help to solve another puzzle which Professor Wood Jones stressed so forcibly, that of understanding how Man obtained his erect posture, and also kept his hands in the primitive, unspecialized, vertebrate condition; for long periods, the hands could not have been used in support of the body as they are in the modern apes, which have never mastered the complete upright position. The chimpanzee slouches forward with his body partly supported by his long arms and with his hands bent up, to take the weight on the knuckles. Man must have left the trees much earlier; in all the modern apes the length of arm is much longer than that of the leg. In Man it is the reverse. The puzzle is: how in fact did Man come to have the perfect erect posture that he has—enabling him to run with such ease and balance? Some have supposed that he could actually have achieved it by such running, or perhaps by leaping, but this does not seem likely. Let me again quote from Wood Jones, this time from his book *The Hallmarks of Mankind*, 1948, p. 78:

"Almost equal certainty may be attached to the rejection of the possibility that he ever served an apprenticeship as a specialized leaper or a specialized runner in open spaces. But it is by no means so easy to reject the supposition that he commenced his career of bipedal orthograde progression as what might be termed a toddler, somewhat after the fashion followed in some degree by the bears."

It seems indeed possible that his mastery of the erect posture arose by such toddling, like children at the seaside. Wading about, at first paddling and toddling along the shores in the shallows, hunting for shellfish, Man gradually went further and further into deeper water, swimming for a time, but having at intervals to rest—resting with his feet on the bottom and his head out of the surface: in fact, standing erect with the water supporting his weight. He would have to raise his head out of the water to feed; with his hands full of spoil he could do better standing than floating. It seems to me likely that Man learnt to stand erect first in the water and then, as his balance improved, he found he became better equipped for standing up on the shore when he came out, and indeed also for running. He would naturally have to return to the beach to sleep and to get water to drink; actually I imagine him to have spent at least half his time on land.

Tied up with his method of assuming the erect position is the problem of the human hand. Let me again quote from Wood Jones (*ibid.,* p. 80):

"In the first place, it seems to be perfectly clear that the human orthograde habit must have been established so early in the mammalian story that a hand of primitive vertebrate simplicity was preserved, with all its initial potentialities, by reason of its being emancipated from any office of mere bodily support. Perhaps the extreme structural primitiveness of the human hand is a thing that can only be appreciated fully by the comparative anatomist, but some reflection on the subject will convince anyone that its very perfections, which at first sight might appear to be specializations, are all the outcome of its being a hand unaltered for any of the diverse uses to which the manus of most of the "lower" mammals is put. Man's primitive hand must have been set free to perform the functions that it now subserves at a period very early indeed in the mammalian story."

Man's hand has all the characters of a sensitive, exploring device, continually feeling with its tentacle-like fingers over the sea-bed: using them to clutch hold of crabs and other crustaceans, to prise out bivalves from the sand and to break them open, to turn over stones to find the worms and other creatures sheltering underneath. There are fish which have finger-like processes on their fins, such as the gurnards; they are just such sensitive feeding organs, hunting for food, and they, too, have been known to turn over stones with them while looking for it.

It seems likely that Man learnt his tool-making on the shore. One of the few non-human mammals to use a tool is the Californian sea otter, which dives to the bottom, brings up a large sea-urchin in one hand and a stone in the other, and then, while it floats on its back at the surface, breaks the sea-urchin against its chest with the stone, and swallows the rich contents. Man no doubt first saw the possibilities of using stones, lying ready at hand on the beach, to crack open the enshelled "packages" of food which were otherwise tantalizingly out of his reach; so in far-off days he smashed the shells of the sea-urchins and crushed lobsters' claws to get out the delicacies that we so much enjoy today. From the use of such natural stones it was but a step to split flints into more efficient tools and then into instruments for the chase. Having done this, and learnt how to strike together flints to make fires, perhaps with dried seaweed, on the sea-shore, Man, now erect and a fast runner, was equipped for the conquest of the continents, the vast open spaces with their herds of grazing game. Whilst he became a great hunter, we know from the middens of mesolithic Man that shell fish for long remained a favorite food.

In such a brief statement I cannot deal with all the aspects of the subject; I shall later do so at greater length and in more detail in a full-scale study of the problem. I will just here mention one more point. The students of the fossil record have for so long been perturbed by the apparent sudden appearance of Man. Where are the fossil remains that linked the Hominidae with their more ape-like ancestors? The recent finds in South Africa of *Australopithecus* seem to carry us a good step nearer to our common origin with the ape stock, but before then there is a gap. Is it possible that the gap is due to the period when Man struggled and died in the sea? Perhaps his remains became the food of powerful sea creatures which crushed his bones out of recognition; or could his bones have been dissolved, eroded away in the tropical seas? Perhaps, in time, some expedition to investigate tropical Pliocene (coastal) deposits may yet reveal these missing links.

It is interesting to note that the Miocene fossil *Proconsul,* which may perhaps represent approximately the kind of ape giving rise to the human stock, has an arm and hand of a very unspecialized form: much more human than that of the modern ape. It is in the gap of some ten million years, or more, between *Proconsul* and *Australopithecus* that I suppose Man to have been cradled in the sea.

My thesis is, of course, only a speculation—an hypothesis to be discussed and tested against further lines of evidence. Such ideas are useful only if they stimulate fresh inquiries which may bring us nearer to the truth.

Statement 2

This is the text of a broadcast talk on the Third Programme (as it was then called) which was published in *The Listener* of May 12, 1960, under the title "Has Man an Aquatic Past?" As most of it is a repetition, in slightly different words, of what appeared in *The New Scientist* article just quoted, only a few short paragraphs of somewhat different material are here reproduced.

... Many animals can swim at the surface if they are forced to, but few terrestrial animals can swim below the surface as man can, or can gracefully turn this way and that to pick up what he is looking for. Native boys diving for coins in a foreign port do indeed look as if they were truly aquatic animals. . . !

Several people have asked me why, if Man has had a long enough evolution in the water to produce such characters as loss of hair and subcutaneous fat, he has not also got webbed hands and feet. Regarding the development of hands, I am sure that selection would not favor such mutations, for his separated fingers would be of greater value in finding and dealing with marine food. But regarding the feet, the truth is that

some people have their toes webbed but they do not like to talk about it!
In 1926, Basler examined 1,000 schoolchildren and found that 9 percent
of boys and 6.6 percent of girls had webbing between the second and
third toes; and in some the webbing may extend between them all. But
apart from the toes the whole foot of Man differs from that of the ape in
that the big toe is joined to the others. This connection is absent in the
apes. It looks as if the human foot may have gone a little way in the
direction of webbing but was later modified for running, and you will
remember that I have supposed that Man was only partly aquatic and,
for at least part of the time, would be walking on the shore.

Students of the fossil record have long been puzzled at the sudden
appearance of Man. The earliest fossil Man, the *Australopithecus,* is
definitely Man, as is shown by the pelvic girdle which is human and not
ape-like: he must have had the erect posture. Before that, there is a great
gap in time, right back to the fossil ape, *Proconsul,* in the middle of the
Miocene. Throughout the whole Pliocene no human remains have been
found, unless we include the doubtful *Oreopithecus.* I would suggest that
perhaps this gap represents the period when Man struggled and died in
the sea. Perhaps there are no coastal tropical *Pliocene* deposits available;
they may have been submerged below the sea.

I have been speaking all the time as if Man was only marine in this
semi-aquatic condition, but he may well have also invaded the rivers,
lakes, and swamps, and so we may yet find his remains in such circumstances. The remains of *Australopithecus* were found in caves, but, not
far from the caves, there are said to be deposits indicating dried-up lakes
or inland seas; so perhaps *Australopithecus* himself was still associated
with water.

Statement 3

This is a popular and somewhat light-hearted article written for *Zenith,*
the magazine of the Oxford University Scientific Society, which is
mainly an undergraduate concern; it appeared in 1977 in Vol. 15, No. 1,
pp. 4-6, under the title "Was there a *Homo aquaticus?*"

Again, a good deal of this article is covering the same ground as that written for the *New Scientist* given above in full under Statement 1.

The following are selected passages that either add some new concepts or give greater emphasis to certain points that were only briefly touched upon in the earlier communications; they represent the author's most recent expression of his views.

. . . Whilst there can be little doubt that man is descended from arboreal ancestors, it is also certain that he came down from the trees at a very early period before his arms became too highly specialized for swinging from bough to bough; he came to feed on the ground. Now here is another important difference between man and the rest of the primates: the latter are essentially vegetarian feeders, living largely on fruits, but with one exception; man alone became a carnivore—the exception being a monkey, the so-called crab-eating macaque, which is now doing just what I believe man did so long ago, going out onto the shores and actually swimming to collect crabs and other crustaceans for food.

We know that man's immediate ancestors were hunting on the land in packs with a leader, like hunting dogs or wolves, and for a time they were very largely carnivorous; the semi-aquatic phase I am envisaging took place long before this. It was here, I believe, that man made that remarkable transition from a fruit-eating diet to one of flesh. How like fruits were the succulent bi-valves that he collected as the tide went out! But that was only the beginning. He became a shellfish eater on a grand scale, and not only of molluscs but crustacea and many other creatures. Competition for food sent him further and further out into the water picking up food from the sea-bed. It was here that he learned to stand upright. We see the same thing happening in the behavior of monkeys in Japan being trained to feed in the sea—they do indeed adopt the erect posture, the water giving their bodies support; man first groped for food on the bottom in shallow water, but stood up to eat it.

The human hand is a remarkable piece of equipment for the picking up of objects between thumb and forefinger (fig. A) and also adapted, I believe, for groping for and seizing living food on the sea-bed. A

Fig. A

mammal that has remarkable human-like fingers on its fore-limbs is the American raccoon which habitually sits by the edge of a stream with its hands in the water feeling about for crayfish or other prey on the bottom. Thus I believe natural selection developed man's remarkable hands, combining the forceps-like finger and thumb for picking up small objects, together with a trap-like cage of fingers for capturing fish and other moving prey. So he went further and further out to sea, swimming from one good fishing ground to another.

. . . We can easily see how natural selection could lead to the reduction of hair for it is reported that the Sydney University Swimming team

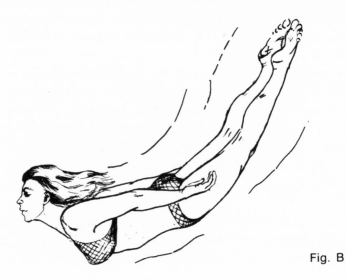

Fig. B

shave off all their body hair before a race and by this save a second in a hundred-yard swim; as groups of our ancestors swam in the tropical seas, chased by sharks, it was the more hairy that tended to lag behind and so become a prey to the voracious jaws. Gradually hair was eliminated except on the head and under the arm pits and in the region of the groin; ladies who for aesthetic reasons shave the hair from their arm pits, suffer considerable discomfort when bathing in that skin tends to rub or stick together, because they have removed the cushion of hair which nature left to prevent this at the junction of limb and body.

Now look at the remarkable stream-line shape of the human form in fig. B; how different from any other of the primates are the beautiful curves of the body helped incidentally by the layers of subcutaneous fat—they are like the curves of a boat, so loved by many men. The rounding of the human jaw, fig. C, unique among the primates, has always been a puzzle to anatomists: it is shaped like the jaws of a frog.

I think it likely that man began to use stones for breaking open the shell-fish, etc., as does the Californian sea otter; and stones are so readily available on the shore. Now let us imagine that on a particular shore man was hammering with a stone and he suddenly found the stone split into thin flakes—flakes of flint—one could almost imagine him crying out with excitement: "Boys—a knife!" but of course he could not speak in that distant age, nor would he know what a knife was, but he could at once see the great advantage of these sharp blades of flint. He began not

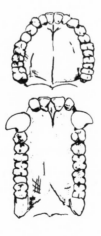

Fig. C

just to use any old stone but to make stone tools like knives and spear heads. He now began to hunt larger marine creatures, spearing large fish, which he could not have caught before, then perhaps even porpoises. So he became a hunter in the sea. Then once he had got his skill and the implements to make it possible, he looked toward the herds of deer and antelopes grazing on the land and he realized that he now had the means of obtaining food in greater quantity, and without all the discomfort of hunting in the sea.

So after some twenty million years or more of living a semi-aquatic life—I must make it clear that I do not suppose man spent more than perhaps five or six hours in the water at a time—*Homo aquaticus* left the sea (or lake) a very different creature from when he first entered it. Now with a hairless body, subcutaneous fat giving him a shapely form, a knowledge of making and using tools, and, above all, the erect posture, he might well be called a new species of man: indeed the ancestor of *Homo erectus*. His feet have always been a compromise between swimming organs and those adapted for running. About this time, I imagine, in fashioning flints he saw the sparks fly which led him to make fires of dried seaweed and driftwood along the beach; he was now equipped to cook the fish he caught.

Perhaps it was not only a shortage of food that sent man to the water in the first place, but also a means of escaping from powerful predators: possibly *Homo aquaticus* was only able to survive and evolve with the help of a number of small sandy or rocky islands stretching up the tropical coasts or margins of lakes where he could live in large colonies, like those of seals or penguins, and where his only enemies were sharks and killer whales in the sea or crocodiles in lakes and rivers.

The only previous publication of my hypothesis was my article in the *New Scientist* of April 1960, and only then was I forced to publish it to protect myself from the outrageous distortions of my views that appeared unexpectedly in the national press. For thirty years I kept the idea to myself, always waiting for the fossil evidence which I felt must surely come. In March 1960, however, I was invited to address a big conference organized by the British Sub-Aqua Club in Brighton and I thought it might be an appropriate moment to try out my ideas, imagining that it would not be reported further than the local *Brighton Argus*. I had not realized that the press of the world was there. My speech was on the Friday evening. Almost every Sunday newspaper came out with banner head-lines such as "Oxford Professor says man is a sea ape!"; some, like the *Sunday Times* and *The Observer*, gave a reasonable summary of my views, but most others were wildly inaccurate. To illustrate a point I had naturally been talking about aquatic mammals like the dolphin, so one paper excitedly declared "Professor Hardy's startling new theory shows man to be descended from a dolphin." I hardly dared to go back to Oxford on the Monday. However, I telephoned the editor of the *New Scientist* to ask if they would publish a more reasonable account of my hypothesis: it came out a fortnight later. I was then asked to give a talk on Radio 3 which was published in *The Listener*. Apart from that I have published nothing further. Desmond Morris devoted a page or two to my ideas in *The Naked Ape* in 1967. He very nearly came down in favor of it, but then decided otherwise, although he went on to say:

"Even if eventually it does turn out to be true, it will not clash seriously with the general picture of the hunting ape's evolution out of a ground ape. It will simply mean that the ground ape went through a rather salutary 'christening ceremony'."

That discussion by Desmond Morris triggered off that well known and witty writer, a former Oxford (Lady Margaret Hall) scholar, Elaine Morgan, to take up the idea and write a book on it. Morris had given no references in the text to indicate whether the ideas he was discussing were his own or those of other people; he did say in the preface, however, that he was deliberately doing this, as it was a popular book, and all the works from which he had obtained his information were listed at the end of the book, but few indeed could tell which idea was taken from which book. Elaine Morgan thought she was taking up an idea that Morris himself had thought of, and then thrown away: so she wrote to ask if she could quote from him. He replied "It was not my idea at all, it is Alister Hardy's—you should write to him." In passing I may say that Desmond Morris tells me he now thinks it likely that I am right.

Elaine Morgan then went to one after another of my various books, two volumes of *The Open Sea, Great Waters,* and *The Living Stream,* but, of course, found nothing whatever; so she wrote to ask me if it was true that I had published on it, and I sent her the *New Scientist* article. I was at that time myself contemplating a book on the subject, but I was not then ready as I had other work on hand. I said that if she could wait a year or two I could give her much more information. However, she was bound by contracts both in America and this country to complete this book by a certain date, so I gave her my blessing to go ahead; indeed, she had every right to do so, for it was now ten years since I had made my views public. Her book, which was published under the title of *The Descent of Woman,* was a best-seller. It was partly about my hypothesis, but also a good deal about woman's place in evolution. She gave me fullest credit for my ideas, and in addition added some very interesting ones of her own, particularly on the origin of tear glands.

I am still waiting for the fossil evidence, but at 81 I must not wait too long! One of the reasons for my accepting the invitation of the editor of *Zenith* to contribute an article was that by choosing this subject I might perhaps persuade some of those in the Geology Department to organize an undergraduate expedition to dig in Miocene deposits which would have marked a tropical shore line (or lake system) in the hope of bringing back solid fossil evidence for *Homo aquaticus.* Alas, most of such deposits are submerged below the Indian Ocean, but the experts may know of a few spots still available. If competent geologists could really put their finger upon them, I have little doubt that funds could be attracted to launch such a search for the missing link. There is still at least a 20 million year gap between the earliest fossil men and their unspecialized ancestor (*Proconsul* and the like). Let Oxford, and *Zenith* readers, fill the gap!

This would really clinch the matter, but now there has come another discovery which is almost as conclusive as the fossil evidence, or so I believe. It has been found experimentally that man has the remarkable adaptation which is found only among mammals and birds that dive under water. It is called the diving reflex and now solves the puzzle of how sponge and pearl divers can remain below so long. It only happens if a man's *face* is submerged; it won't occur if he wears a mask. If he dives under water and his face exposed, there is an immediate reaction cutting down the blood supply to most of the body, but leaving a good supply to both the brain and the muscles of the heart. This reaction is typical of whales, seals, penguins, and even diving ducks: I cannot believe that it could have been evolved by natural selection unless man had taken to diving under water some considerable period of his past history. The only remaining test to be made is to persuade some physiologists to do simple experiments with all the known apes. They merely have to be put in a bath of water with their faces submerged for a short time whilst an electro-cardiograph records the changes in the circulation of their blood. If man is really unique in this I am home and dry! But in addition it would be very pleasant in my old age to have a bit of fossil *Homo aquaticus*—or a cast of it—on my mantelpiece; so perhaps the Oxford Exploration Club might think of pandering to my eccentricity.

All this, of course, is only an hypothesis and valueless till put to the test. Speculation is the fuel of scientific progress; it drives forward to discovery only if it is continually being burnt in a fire of constructive criticism. Let the critics open fire.

Appendix 3

Neoteny in Man

The basic list of neotenous characteristics in *Homo sapiens* as compiled by Louis Bolk:

(1) Our "flat-faced" orthognathy.

(2) Reduction or lack of body hair.

(3) Loss of pigmentation in skin, eyes, and hair (Bolk argues that black peoples are born with relatively light skin, while ancestral primates are as dark at birth as ever).

(4) The form of the external ear.

(5) The epicanthic (or Mongolian) eyefold.

(6) The central position of the foramen magnum (it migrates backward during the ontogeny of primates).

(7) High relative brain weight.

(8) Persistance of the cranial sutures to an advanced age.

(9) The labia majora of women.

(10) The structure of the hand and foot.

(11) The form of the pelvis.

(12) The ventrally directed position of the sexual canal in women.

(13) Certain variations of the tooth row and cranial sutures.

Appendix 4

The Elephant

The concept of once-terrestrial animals having returned to the sea is a very familiar one. The concept that an animal, having once made that transition, might subsequently return to the land and readapt to terrestrial life is unfamiliar, but in evolutionary terms there is nothing intrinsically unacceptable about it.

There is one other animal that bears many of the hallmarks of having made the double transition—very much earlier than the aquatic ape. The aquatic theory does not stand or fall by the proposition that the elephant is another example of an ex-aquatic animal, but there are many features of his physiology and behavior which would support such a hypothesis.

    (1) Like many aquatic mammals he is virtually hairless, except for a tuft on his tail and sometimes a patch of hair on his head. Some extinct forms, such as the mammoth which inhabited mainly sub-Arctic environments, possessed a hairy coat; and the young of the present-day elephants are born covered with yellow and brown hair which is later shed—possibly analogous to the lanuginous coat of the human fetus.

    (2) Elephants have webbing between their toes.

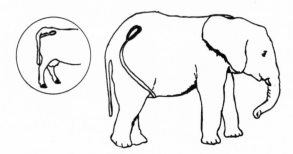

Fig 21.   The elephant's vaginal canal follows an extraordinary route.

(3)   The vaginal canal of the female elephant follows a route unknown in any other terrestrial mammal (see fig. 23). It emerges in such an unusual position that it used to be believed that elephants copulated ventro-ventrally.

(4)   Carcases of mammoths have more than once been found in the Arctic completely preserved by the ice. A subcutaneous layer of fat about 8 centimeters thick covered the whole body.

(5)   The male elephant's sex organs are streamlined. The testicles, like those of the whale, are externally invisible, lying within the body wall.

(6)   The opening in the skull for the nostrils has migrated dorsally, as with whales and other marine mammals: it emerges above the eyes. This is not readily observable because in the soft tissue the air canal thereafter continues downward and emerges at the end of the trunk.

(7)   The diaphragm is oblique, as in whales and dugongs.

(8)   The elephant is an excellent swimmer. He has been known of his own volition to swim distances of up to 300 miles to offshore islands.

(9)   He sheds tears when emotionally disturbed.

(10)   He has voluntary control of vocal utterances: elephants can easily be trained to trumpet to order.

(11)   When an elephant gives birth, another female (popularly described as a "midwife elephant") stands by until the process is complete. The

Encyclopaedia Brittanica attributes this to defense against predators. "In regions where large carnivores such as tigers prey upon newborn elephants, the cow seeks a female associate." However, it would take a very brave or hungry lion to approach a 5-ton elephant in labor closely enough to snatch away her baby. The "female associate" is strongly reminiscent of the dolphin midwife who stays near to assist the newborn to the surface for its first breath. This behavior is known in no other terrestrial species, with the exception of our own.

The evolution of tusks has never been satisfactorily explained. They occur in several mammalian species which are totally unrelated to one another. These species include—apart from the elephant and its extinct relatives—the walrus, the sea cow, and the babirusa. All have naked hides, and all are either aquatics or swamp dwellers.

Bibliography

ADOLPH, E. F. (and associates), *Physiology of Man in the Desert*, Interscience Publishers, Inc., 1947.

BERTRAM, C., *In Search of Mermaids: The Manatees of New Guinea*, Peter Davies, London, 1963.

CARRINGTON, RICHARD, *Elephants: A Short Account of their Natural History, Evolution and Influence on Mankind*, Chatto & Windus, London, 1958.

CHANCE, MICHAEL and JOLLY, CLIFFORD. *Social Groups of Monkeys, Apes and Men*, Jonathon Cape, London, 1970.

COUSTEAU, JACQUES, *The Ocean World of Jacques Cousteau: Mammals in the Sea*, Angus & Robertson, London, 1980.

DARWIN, CHARLES, *Descent of Man*, Random House, Inc., New York, 1871.

DARWIN, CHARLES, *The Expression of Emotions in Man and Animals*, University of Chicago, Chicago, 1965.

DOYLE, G. A. and MARTIN, R. D. (eds.), *The Study of Prosimian Behavior*, Academic Press, New York, 1979.

FANGE, R., SCHMIDT-NIELSEN, K., and OSAKI, H., "The salt gland of the herring gull," *Biological Bulletin*, 115:162. 1958.

FICHTELIUS, KARL-ERIK and SJOLANDER, SVERRE, *Man's Place: Intelligence in Whales, Dolphins, and Humans*, translated by Thomas Teal, Gollanc, London, 1973.

FRINGS, H., ANTHONY, A., and SCHEIN, M. W., "Salt excretion by nasal gland of Laysan and Blackfooted albatrosses," *Science*, 128:1572, 1958.

GOULD, STEPHEN J., *Ontogeny and Phylogeny*, Harvard University Press, Harvard, 1980.

GOULD, STEPHEN J., *The Panda's Thumb*, W. W. Norton & Co., New York, 1980.

HARDY, ALISTER, *The Open Sea*, Collins, London, 1956.

HARLOW, HARRY F. and MARGARET K., "Social Deprivation in Monkeys," *Scientific American*, 1962.

HUXLEY, JULIAN, *Evolution: The Modern Synthesis*, Allen & Unwin, London, 1963.

IRVING, L., "Respiration in diving mammals," *Physiological Reviews*, 19: 112, 1939.

JERISON, H. J., *Evolution of the Brain and Intelligence.* Academic Press, New York, 1973.

JONES, FREDERIC WOOD, *Man's Place Among the Mammals*, Edward Arnold, London, 1929.

KELLOGG, W. N., *Porpoises and Sonar*, University of Chicago Press, Chicago, 1961.

LEAKEY, MARY D., "Ashes of Time," *National Geographic Magazine*, April, 1979.

LEAKEY, RICHARD E. and LEWIN, ROGER, *Origins*, MacDonald & Jane's, London, 1977.

LEE, RICHARD B. and DE VORE, I., *Man the Hunter*, Aldine Publishing Co., 1968.

LE GROS CLARK, W. E., *History of the Primates*, University of Chicago Press, Chicago, 1965.

LIEBERMAN, PHILIP, *On the Origins of Language: An introduction to the evolution of human speech*, Macmillan, 1975.

LOCKLEY, R. M., *Grey Seal, Common Seal*, André Deutsch, London, 1966.

LOVEJOY, C. OWEN, "The Origin of Man," *Science,* 23 January, 1981.

MARTIN, R. D., "Phylogenetic aspects of prosimian behavior" in *The Study of Prosimian Behavior,* Academic Press, London, 1979.

MARTIN, R. D., "Adaptation and body size in primates," *Z. Morph. Anthrop,* 71, 2, 115-124. August, 1980.

MARTIN, RICHARD MARK, *Mammals of the Sea,* B. T. Batsford Ltd., London, 1977.

PFEIFFER, JOHN E., *The Emergence of Man,* Thomas Nelson, London, 1970.

RUDDER, BENJAMIN CHARLES COLLYER, The Allometry of Primate Reproductive Parameters, Ph.D. Thesis, London, 1980.

ROE, ANNE and SIMPSON, GEORGE GAYLORD (eds.), *Behavior and Evolution,* Yale University Press, 1969.

SCHEFFER, VICTOR B., *Seals Sealions and Walruses,* Oxford University Press, Oxford, 1958.

SCHMIDT-NIELSEN, K. and FÄNGE, R., "Salt glands in marine reptiles," *Nature,* 182: 783. 1958.

SMITH, HOMER W., *From Fish to Philosopher,* Doubleday & Co., 1961.

LAWICK-GOODALL, JANE VAN, *The Shadow of Man,* Collins, London, 1971.

VIRGO, H. B. and WATERHOUSE, M. J. "The emergence of attention structure among rhesus macaques," *Man,* vol. 4, no. 1, March 1969.

WIND, JAN, "Human Drowning: Phylogenetic Origin," *Journal of Human Evolution* (1976), 5, 349-63.

ZIHLMAN, A. and LOWENSTEIN, R., "False Start of the Human Parade," *Natural History,* Aug.-Sept., 1979.

YAPP, W. B., *Vertibrates, Their Structure and Life,* Oxford University Press, New York, 1965.

YOUNG J. Z., *The Life of Vertebrates,* Clarendon Press, 1962.

WASHBURN, SHERWOOD L. "Tools and Human Evolution" in *Human Variation and Origins,* W. H. Freeman & Co., 1967.

WILLIAMS, LEONARD, *Man and Monkey,* Deutsch, London, 1967.

Index